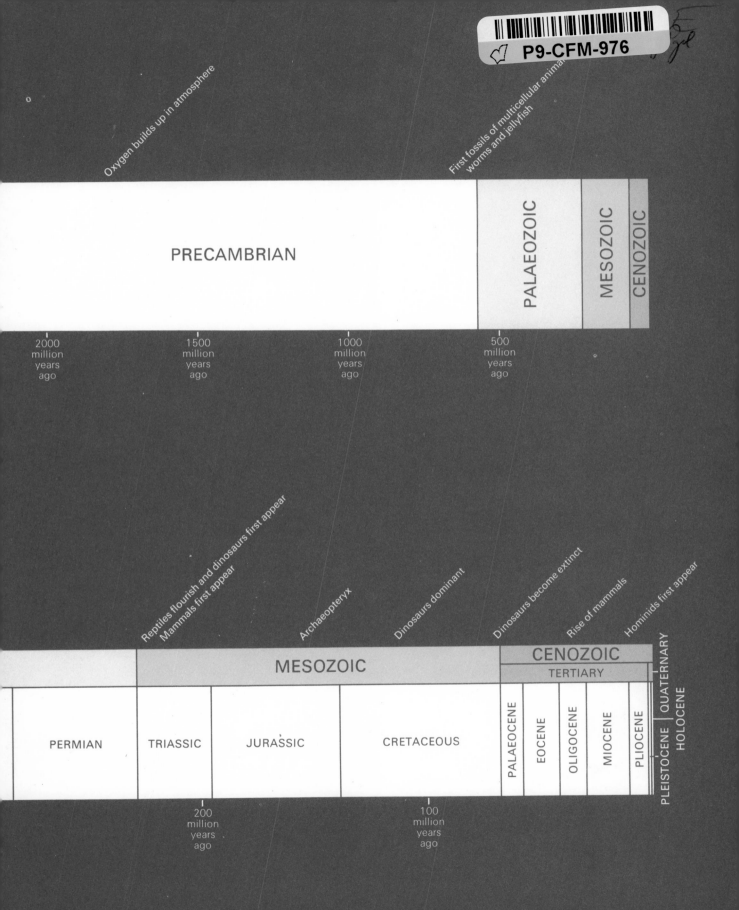

Oxygen builds up in atmosphere

First fossils of multicellular animals worms and jellyfish

PRECAMBRIAN

PALAEOZOIC

MESOZOIC

CENOZOIC

2000 million years ago

1500 million years ago

1000 million years ago

500 million years ago

Reptiles flourish and dinosaurs first appear

Mammals first appear

Archaeopteryx

Dinosaurs dominant

Dinosaurs become extinct

Rise of mammals

Hominids first appear

MESOZOIC

CENOZOIC

TERTIARY

QUATERNARY

PERMIAN

TRIASSIC

JURASSIC

CRETACEOUS

PALAEOCENE

EOCENE

OLIGOCENE

MIOCENE

PLIOCENE

PLEISTOCENE

HOLOCENE

200 million years ago

100 million years ago

The Illustrated
ORIGIN OF SPECIES

The Illustrated

ORIGIN OF SPECIES

=== *By* ===

Charles Darwin

Abridged & Introduced by

RICHARD E. LEAKEY

Consultants

W. F. BYNUM M.D., Ph.D.
University College, University of London, England

J. A. BARRETT Ph.D.
University of Cambridge, England

Ⱡⱳ

Hill and Wang · New York
A division of Farrar, Straus and Giroux

First published in the United States of America 1979 by
Hill and Wang, a division of Farrar, Straus and Giroux
Published simultaneously in Canada by
McGraw-Hill Ryerson Ltd, Toronto

ISBN 0-8090-5735-2

This book was designed and produced by
The Rainbird Publishing Group Ltd
36 Park Street, London W1Y 4DE

House Editors: Linda Gamlin, Michael Scherk
Designer: Yvonne Dedman
Picture Researcher: Elizabeth Eyres
Production: Georgina Ewer

The text was set by SX Composing Ltd, Rayleigh, Essex, England

The colour plates were originated and the books printed and
bound by Dai Nippon Printing Company Ltd, Tokyo, Japan

Library of Congress Cataloging in Publication Data

Darwin, Charles Robert, 1809–1882
The illustrated origin of species.

Bibliography: p. 240. Includes index.
1. Evolution.   2. Natural selection.
I. Leakey, Richard E.   II. Title.
QH365.02   1979   575   79-50923
ISBN 0-8090-5735-2

Contents

Acknowledgments

In reading *The Origin of Species* one cannot fail to be struck by the enormous range of subjects which Darwin traversed in search of evidence for his theories. Today it is impossible for one man to be fully acquainted with all the relevant details of such diverse subjects, and I am grateful for the kind help received from a number of people, each of them an expert in a particular field. I am especially indebted to the two consultants, Dr William Bynum and Dr John Barrett, whose breadth of knowledge has been invaluable. I would also like to acknowledge with thanks the advice of the following people: Dr John Thackeray, Mr Peter Gautrey, Dr Philip Burton, Mr John Marchant, Mr J. E. Hill, Dr Gordon Corbet, Dr Humphry Greenwood, Dr Bruce Campbell and Dr Paul Whalley. Lastly, I would like to offer my special thanks to Linda Gamlin who has been such a help in putting together this very complex book.

Introduction
by Richard E. Leakey

Darwin and his theory

The entire first edition of *The Origin of Species* was sold out on its publication day, 24 November 1859. A second edition was ready by January 1860, and the book went through a total of six editions during Darwin's lifetime. It has been constantly in print since its first appearance and has been translated into some thirty languages. Although it is first and foremost a scientific classic, *The Origin of Species* was written for the educated general reader of Darwin's time and as such has a unique literary status among works of the scientific imagination. It was an important book in Darwin's time and it remains so today, for the theory of evolution is the cornerstone of modern biology and Darwin's book forms the foundation of that theory.

Nevertheless, Darwin was not the first person to propose that species of plants and animals can change with time. In the later part of the eighteenth century Charles's grandfather, Erasmus Darwin, wrote an account of evolution, and shortly afterwards, in 1809, the French naturalist Jean Baptiste de Lamarck published his *Philosophie zoologique*, which contained his own speculations on the mutability of biological species. Indeed, Charles Darwin himself noted no less than twenty predecessors who had written on aspects of evolution. Yet it is from Darwin that modern evolutionary theory derives. Why? Two main reasons may be given.

Firstly, Darwin patiently and systematically sifted through all the kinds of evidence bearing on his subject. As a young man he had spent five pro-

ductive years as naturalist on H.M.S. *Beagle* (1831–1836). During this long voyage around the world Darwin developed into an excellent naturalist, constantly observing, collecting, and thinking about the many geological and biological phenomena which confronted him. As early as 1837 he began to doubt that species are permanent and immutable, and though he was busy from 1837 to 1859 with many scientific activities, this question of the origin of species – what scientists of his day called 'the mystery of mysteries' – was frequently before him. During those years he read widely, thought deeply, and experimented carefully. As a result, *The Origin of Species* is a work of remarkable breadth and depth.

Secondly, Darwin was able to provide a plausible mechanism to explain how species can change: *natural selection*. Darwin first hit upon the idea of natural selection in 1838, after reading *An Essay on the Principle of Population*, by Thomas Malthus, a clergyman and political economist of the early nineteenth century. Malthus was primarily concerned with human populations, but he pointed out that it is a general principle of nature that living organisms produce more offspring than can normally be expected to survive to reproductive maturity. An oak produces hundreds of acorns annually, a bird can bring forth several dozen young in its lifetime, and a salmon lays thousands of eggs per year, each of which can potentially become an adult. Despite this massive reproductive capacity, adult populations tend to remain stable from generation to generation.

Malthus's work helped Darwin grasp an important point: there could be selection between the

Charles Darwin in 1881, twenty-two years after *The Origin of Species* was first published.

offspring as to which ones survived and which perished. Since the individual members of a species vary slightly among themselves, those individuals with certain characteristics which give them an advantage, in gaining food or escaping predators for example, will have an enhanced chance of survival. Given variation, the results of which can be observed everywhere, natural selection can, according to Darwin, account for biological evolution (or what he called 'descent with modification'). Just as man can effect dramatic changes in domestic animals by artificially selecting characteristics which he finds desirable, so nature 'selects' those members of a species best able to cope with the rigours of life. As environmental conditions change, natural selection assures that certain characteristics in a randomly varying population are favoured. As the nineteenth-century evolutionary philosopher Herbert Spencer put it, nature guarantees 'the survival of the fittest'.

Darwin, then, was not the first to put forward a theory of evolution, but his was the first mature and persuasive account. As it turns out, even the discovery of the principle of natural selection was not uniquely his, for the naturalist Alfred Russel Wallace (1823–1913) independently discovered it in 1858, before Darwin had published the results of his patient enquiries. Wallace had not met Darwin, but knew of his reputation as an expert naturalist with certain unorthodox opinions, and they had corresponded, in general terms, on the question of whether species were permanent. He sent Darwin his short essay, entitled *On the Tendency of Varieties to Depart Indefinitely from the Original Type*, in which the principles of natural selection were explained, without having any idea that Darwin had already discovered natural selection, or of the extent to which his essay pre-empted Darwin's lifework. In his accompanying letter Wallace requested that Darwin should read the essay and, if he thought it worthwhile, forward it to Charles Lyell. Darwin was initially at a loss as to what he should do, but Charles Lyell and Joseph Hooker,

The naturalist Alfred Russel Wallace. He independently discovered the principle of natural selection in 1858 and this spurred Darwin into publishing his own theories. Wallace always gave Darwin the credit for being the first to discover natural selection. While he firmly believed in evolution, unlike Darwin he maintained that the human mind could not have originated by evolutionary processes.

who were sympathetic to his views and had been urging him to publish for some time, arranged for a joint Darwin–Wallace memoir on natural selection to be read at the Linnean Society in July 1858 and subsequently to be published in the *Journal of the Linnean Society* for 1858. Both Darwin and Wallace acted most generously over the matter, and though Wallace was himself a very distinguished naturalist, he never failed to give Darwin the major credit for first discovering the principle of natural selection. Like Darwin, Wallace had been led to natural selection through reading Malthus. As a result of Wallace's own endeavours, Darwin gave up a long book on natural selection which he had started writing in 1856 (never finishing it), and instead turned his energies to a shorter work on evolution.

This shorter work was *The Origin of Species*. As we have seen Darwin's achievement in that book was two-fold: he marshalled the evidence for evolution, and he described a mechanism by which new species could be formed. In his lifetime,

Darwin was most successful with the first of these contributions, for *The Origin of Species* convinced many of Darwin's open-minded colleagues that biological evolution does occur. So persuasive was the range of evidence which Darwin assembled that, on first reading *The Origin of Species*, Thomas Huxley could only remark, 'How extremely stupid not to have thought of that!' But even Huxley, who called himself 'Darwin's bulldog' and was the most vigorous defender of Darwin's work in the later nineteenth century, did not believe that natural selection had been demonstrated as the primary mechanism of evolutionary change. Because natural selection had not been subjected to experimental proof, Huxley and others withheld wholehearted assent, and even Darwin began to search for additional mechanisms. Darwin wrote another book about one of these accessory mechanisms – sexual selection – and he discussed it briefly in the sixth edition of *The Origin of Species*: the edition which forms the basis of the present volume. But it is clear that even extensive criticism of the adequacy of natural selection to account for evolution could not shake Darwin's conviction that it is by far the most important mechanism. It is only in the twentieth century that we have been able to appreciate fully the genius of Darwin's tenacity, for modern genetics and much field observation and laboratory research have shown us the true explanatory power of Darwin's theories. Especially since the 1920s, with the writings of the population geneticists such as Sir Ronald Fisher, J. B. S. Haldane, and Sewall Wright, it has been apparent that a synthesis of Darwin's work on natural selection with Gregor Mendel's work on genetics results in a cohesive and intelligible picture of evolutionary change. This has been called the *neo-Darwinian synthesis*: Darwinian because it accepts natural selection, 'neo' because it uses theories of heredity which Darwin did not know about, even though Mendel actually published his classic work in 1866.

We shall return to the details of this neo-Dar-

winian synthesis. It is significant that many of these neo-Darwinians of the 1920s were concerned with whole *populations* of plants and animals, for it was above all Darwin who inaugurated the shift in modern biology from typological to population thinking. Before Darwin, most biologists had thought of species as fixed and eternal groups, ordained by God. Since God had directly created the individual species of plants and animals, each species had its essential defining characteristics – tacitly assumed to be those possessed by the original member or reproductive pair of each type or species. Although naturalists recognized that some variation within a species was possible, pre-Darwinian scientists insisted that there were natural limits to this variation.

Darwin showed how 'descent with modification' could result in the appearance of new species. Like modern biologists, he realized that at a fixed

Thomas Huxley was a young man of thirty-three when he read *The Origin of Species* and remarked 'How extremely stupid not to have thought of that'. He became Darwin's most vocal supporter and engaged in bitter public debates with his critics, notably Bishop Wilberforce of Oxford and the distinguished comparative anatomist Richard Owen. Darwin himself shunned these encounters and was glad to leave Huxley to act as his 'bulldog'. Huxley is pictured here in 1882, the year of Darwin's death.

Many of Darwin's colleagues, including Huxley, although convinced that evolution does occur, failed to appreciate the efficacy of natural selection. In an attempt to answer their criticisms Darwin searched for additional mechanisms and one that he correctly identified was sexual selection. In this painting of ruffs by John Gould, sexual selection can be seen in action. The male ruffs in the background are fighting for a dominant position in the 'lek' or mating arena. The outcome of this batttle will in part be determined by the size of the collar of feathers around each male's neck. Once a male has gained such a position the collar of feathers will be displayed to attract the females for mating as shown in the foreground. A male with an unimpressive collar of feathers will be unlikely to find a mate and will leave few or no offspring. There is, therefore, strong selection in favour of larger collars of feathers in the male ruff.

point in time, species do have some real existence: the biological notion is not simply a human construct. For the purposes of his argument Darwin avoided giving an exact definition, but embedded in *The Origin of Species* is the idea of a species close to that held by contemporary scientists, who define it as a population which is simultaneously a reproductive community, an ecological unit, and a genetic unit. This means that the individuals constituting a given species can interbreed one with another; they occupy the same ecological niche, generally compete for the same food, and suffer the same kinds of predators; and they share the same pool of genes which passes from generation to generation. Darwin would recognize most of the

methods a modern biologist uses to differentiate species and to elucidate the processes of species formation.

This should not be taken to imply that evolutionary biology has not made tremendous advances since Darwin's time. It has, and it is the purpose of this edition both to bring Darwin's great work before the modern reader, and to explain how modern research has extended our understanding of the problems which Darwin faced. I have selected the text from the sixth edition of *The Origin of Species*, published in 1872, only ten years before Darwin's death at the age of seventy-three. It therefore embodies Darwin's final thoughts on the subject. The text has been cut to about a third of its original length by eliminating much repetition, some of Darwin's examples, and his tendency to Victorian wordiness. On two or three occasions, I have taken the liberty of slightly rearranging the order of paragraphs to make the condensed text read more smoothly, and in making the cuts I have frequently had to interpolate words or phrases. But these interpolations in no way change Darwin's original meaning. I have not tried to make Darwin more modern than he was, since he can stand on his own feet. Sometimes, though, when subsequent research has shown Darwin's discussions to be wrong, I have added short explanatory notes. Modern names for animals and plants are inserted where the ones Darwin gives are no longer used. If a popular name exists for a species which Darwin refers to only by a Latin name, this is inserted. However there has been no attempt to update the Latin names, many of which have now been superseded. Above all, I agree with Darwin when he insisted that *The Origin of Species* constitutes one long connected argument. I hope that the present volume will make this argument even clearer. *The Origin of Species* is still as good an introduction as any into the question of evolution. Darwin's genius transcends the century separating him from us and makes him simultaneously an historical scientist of great originality and our contemporary.

Darwin's problems

One of the virtues of *The Origin of Species* is the way in which Darwin was not afraid to face the difficulties confronting his theory of descent with modification by natural selection. Because evolutionary biology does not lend itself to the kinds of proof on which disciplines such as chemistry and physiology rely, it is still, to some extent, an inferential science, and in reading *The Origin of Species* we are struck with the frequency with which Darwin worked by assembling bodies of facts and observations explicable by his theory but difficult to explain by the conventional assumptions of his time. Darwin's theory has beauty and grandeur because it explains so much so economically, and it is not surprising that the status of evolution by natural selection has often been compared to Newton's theory of universal gravitation, which bears equally on the fall of an apple, the movements of the tides, and the orbit of a planet around the sun.

Nevertheless Darwin did have problems, many of which he recognized and dealt with in the first edition of *The Origin of Species*. After 1859, critics raised others and in subsequent editions Darwin methodically considered these. In most cases Darwin was able to provide completely adequate

explanations of miscellaneous anomalies and difficulties, and these can be read in the text. But it might be helpful here to mention three or four major obstacles with which Darwin had to grapple. In this section we are primarily concerned with the historical Darwin, but I shall briefly mention some recent developments, reserving a more extended discussion of modern aspects of evolutionary biology for the final section.

We can begin with one problem which initially did not particularly concern Darwin: time. In the early decades of the nineteenth century the literal Biblical chronology of earth history had been seriously challenged. Through the work of geologists and palaeontologists such as Georges Cuvier in France and William Smith and Charles Lyell in Britain, scientists and educated laymen alike became aware of the immensity of geological time. They learned that the earth had previously been inhabited by species of plants and animals now extinct: in particular the Victorian imagination, like our own, warmed to the romanticism of the huge dinosaurs and other extinct beasts, whose remains could be found even in Britain. On New Year's Eve 1853 Richard Owen (later to be one of Darwin's most ardent critics) and others dined in splendour inside a massive model of a dinosaur. Even in 1831, when Darwin left on the *Beagle*, he knew that the geological phenomena he would witness would not fit into the 6,000-year timescale calculated on a literal reading of the Bible. Charles Lyell's *Principles of Geology* was particularly important to Darwin. The first volume was published in late 1830, just before he left England: volume two reached him in South America. Lyell set himself the task of explaining earth history only by referring to causes still in operation. In particular these causes included land elevation by volcanic and earthquake action and erosion by the effects of wind and water. Since the phenomena to be elucidated included not only monumental mountain ranges such as the Alps and Andes, but the existence of marine fossils in strata

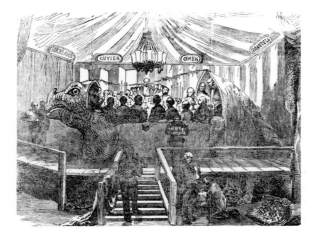

Richard Owen and twenty-one other noted scientists dined inside a massive model of a dinosaur in the grounds of the Crystal Palace, London, on New Year's Eve 1853. The discovery of fossils of extinct creatures such as dinosaurs had already challenged the literal acceptance of the Biblical creation story, but it was possible to reconcile the two by supposing that there had been several successive creations followed by mass extinctions, of which Genesis dealt only with the last.

found on mountain tops, it was apparent that vast reaches of time were necessary for the success of the Lyellian programme.

The explanatory elegance of Lyell's work impressed Darwin, for its success seemed to guarantee that time presented no problem. However, the fossil record was a different matter. One of the discoveries on which the time revolution of the early nineteenth century rested was the recognition that particular kinds of fossilized plants and animals were associated with particular geological strata. Fossils could thus be used as aids in the relative dating of strata. As the twin sciences of geology and palaeontology matured from the 1820s, stratigraphical observations from many parts of the globe began to be placed together to yield a coherent picture of the history of life on earth.

But while the sequence of fossils seemed to possess a directional quality, with simpler plants and animals appearing before more complicated forms, and the fossils of more recent strata bearing a closer resemblance to living species, geologists

before Darwin did not place an evolutionary interpretation on their findings. There were three main scientific reasons for this. In the first place, the earliest know fossils were relatively complicated animals – mostly marine invertebrates such as molluscs and crustaceans. Geologists of the 1830s were convinced that they had discovered the dawn of life in this strata, called the Cambrian, and no hypothesis save special creation was deemed adequate to account for the sudden appearance of Cambrian fossils. Rocks older than the Cambrian were, so it seemed, completely devoid of fossils. In the second place, the various strata each generally had its own characteristic fossil flora and fauna and the transitions between strata were abrupt. This suggested to geologists the probability of successive wholesale creations and extinctions. Indeed, one common way of reconciling Genesis and geology was to assume that the Biblical story referred only to the final creation of present plants and animals, including, of course, man himself. There was no reason, it was argued, for the story of the earlier epochs to have been included in Genesis, since they did not concern man's spiritual needs. Even in the 1820s, there were serious problems with this compromise, for the discovery of human remains in English caves could be (and later was) interpreted as proving the coexistence of man with various extinct species of mammals, such as mammoths. But until the late 1850s, man was almost universally assumed to be of recent origin. When the evidence favouring a great antiquity for man became irrefutable by the early 1860s, it subtly helped Darwin's case, though he never exploited it in *The Origin of Species*, believing that a too extensive treatment of man would prejudice scientists against more general aspects of his theory of descent with modification. Darwin himself held that man is the product of evolution, as his early private notebooks on evolution from the late 1830s testify. He considered the case of human evolution in two books published after 1859, *The Descent of Man and Selection in Relation to Sex* (1871)

One of the main objections made to Darwin's theory was that no fossils had been found of animals intermediate between the major vertebrate groups. The finding, in 1861, of Archaeopteryx, intermediate between reptiles and birds, was a triumph for Darwin. Archaeopteryx possessed certain birdlike features, notably feathers, but like a reptile it had teeth and a long tail. Anatomical studies show that its muscles would not have been strong enough to enable it to fly, and that it must have glided from tree to tree. It probably climbed up the trees with the aid of the claws on its wings.

and *The Expression of the Emotions in Man and Animals* (1872).

Religious considerations undoubtedly played a role in the refusal of geologists to place an evolutionary interpretation on their palaeontological discoveries. But in addition to the geologically sudden appearance of signs of life in the Cambrian strata and the abrupt transitions between contiguous strata, there was another closely related reason why individual creation rather than gradual evolution was the dominant scientific assumption before *The Origin of Species*. This involved the dramatic appearance in the fossil record of some major groups, and in particular of the various classes of vertebrates (fish, amphibians, reptiles, birds, mammals). Few transitional fossils seemed to exist, and physiological reasoning suggested that there was no conceivable gradual path from gills to lungs, or from a normal vertebrate forelimb to a wing. Various evolutionists before Darwin, such as Geoffrey St. Hilaire in France and Robert Chambers in Britain, developed evolutionary hypotheses which relied on the assumption that abrupt changes could occur during embryological development, but most scientists believed that this merely substituted one kind of miracle for another, and continued to believe in special creation. This problem was particularly acute for Darwin, for he always insisted that evolution was gradual rather than saltatory, and yet he too was faced with a fossil record which seemed to bespeak sudden changes.

The origin of major groups is still a matter of some dispute among palaeontologists, though new fossil finds since Darwin's day have smoothed out much of the abruptness of the fossil record. Even in Darwin's lifetime certain transitional forms were found, such as Archaeopteryx, intermediate between reptiles and birds. Other fossil finds have closed the gaps between fish and land vertebrates, and between reptiles and mammals. Many palaeontologists today believe that evolution has always proceeded at the same steady rate and that the absence of transitional forms can be explained by the arguments which Darwin himself used, particularly that there are immense gaps in the fossil record and that transition usually occurred in one restricted locality, with migration from that locality only taking place when the new form had reached a fairly advanced stage. But some palaeontologists, notably Stephen Jay Gould of Harvard, think the fossil evidence suggests that at various stages in the history of life evolution has progressed unusually rapidly – in 'spurts' – and that major branching in the evolutionary tree has occurred at these points. This may have occurred because an evolving group had reached a stage in its organization at which it had an advantage over other groups and, once this threshold had been passed, it was able to radiate rapidly to exploit a variety of niches. Or a 'spurt' may have followed from a widespread change of climate, or from the extinction of another group of species leaving many niches vacant.

Related to the evolution of major groups is the more general problem of complex organs, whose gradual development is difficult to conceive if only the fully functional organ is of service to the individual. How can an evolutionary line acquire feathers for flight if rudimentary feathers would be useless? Darwin himself was particularly concerned with the problems associated with the eye, since that organ was frequently cited by creationists as proof that God had directly fashioned every individual species. As Darwin confessed in 1860 to the American naturalist Asa Gray, 'The eye to this day gives me a cold shudder, but when I think of the fine known gradations, my reason tells me I ought to conquer the cold shudder.' Darwin's approach to this problem was in general like that of modern evolutionary biologists – a combination of comparative anatomy, embryology, physiology, palaeontology, and reason.

Despite these issues, Darwin was correct to argue that the fossil record holds no insurmountable problems for evolutionary theory, and his

The naturalist Asa Gray. He and Darwin met twice in London and corresponded from 1855 until Darwin's death. A letter to Asa Gray from Darwin, written in 1857, which outlined his theory of descent with modification, was incorporated into the Darwin-Wallace memoir of 1858, as proof of the fact that Darwin had been thinking about natural selection for several years before Wallace had independently discovered the same principle. Asa Gray was professor of botany at Harvard and the major champion of Darwinism in America. His principal opponent was Louis Agassiz, a fellow professor at Harvard.

discussion in Chapters 10 and 11 of *The Origin of Species* may still be read with profit. As we shall see in the next section, micropalaeontology has, within the last decade or two, vastly extended the history of life on earth in ways which could only have pleased Darwin. In a sense, modern palaeontology, with its new discoveries and new techniques such as radioactive dating, has simply confirmed his theories. But in the case of another of his problems – that of heredity – modern science has been able to go much further than Darwin.

Darwin did not know the mechanism of heredity, but he knew that variation does occur and that some variation at least is hereditary. He was particularly interested in the relatively wide variations that can be seen in domesticated plants and animals, not only because of the analogy he drew between artificial and natural selection, but also because domesticated plants and animals seemed to him to offer the greatest hope for understanding the more general phenomena of inheritance.

Darwin himself performed breeding experiments with various domesticated species, particularly pigeons. (The pigeon figures frequently in *The Origin of Species* and when Darwin's publisher submitted the manuscript of the book to a referee for an opinion, the referee wrote back regretting that Darwin had not simply written a book on pigeon-breeding: 'Everybody is interested in pigeons', he insisted. 'The book would be reviewed in every journal in the kingdom, and would soon be on every library table.') On the basis of his experiments with domesticated species, Darwin wrote a two-volume treatise on domestication, published in 1868, in which he developed a theory of heredity which he called 'pangenesis'.

In this theory Darwin postulated that each organ of the body gives off microscopic particles which he called 'gemmules' and that these gemmules are collected in the reproductive organs where they can influence the corresponding shape, size, and functions of the organs of the offspring. Darwin wanted to account for his belief that some characteristics or habits acquired during a parent's lifetime could affect the corresponding attributes of its offspring. This he usually called 'use-and-disuse heredity'; the more familiar phrase is 'the inheritance of acquired characters' or 'Lamarckism', named after Jean Baptiste de Lamarck. However Lamarck's theory was rather different in that he additionally postulated a desire for change, or *besoin,* which caused that change to happen in the organism itself and then to be passed on to its offspring. In Darwin's time this aspect of Lamarck's theory was not generally accepted, but virtually every scientist believed that characters acquired by use or disuse could be inherited.

Darwin developed his theory of pangenesis in order to describe how use-and-disuse inheritance could occur. But he realized that his theory was purely speculative and therefore put little faith in it. When it was soundly criticized by his cousin Francis Galton (himself an eminent psychologist, statistician and student of heredity), Darwin quietly dropped pangenesis, though he continued to accept the inheritance of acquired characters. More generally, however, Darwin correctly insisted that natural selection required only that hereditary variation occur. Natural selection thus did not imply any particular mechanism of heredity and this is the reason why Darwin's evolutionary theories have been so easily integrated with modern genetics. Since Darwin knew that he did not understand hereditary mechanisms he tried to confine himself to observations at what we would now describe as the level of *phenotypic* variation (*see the diagram on p. 18 for the meaning of the term phenotype*). Many of the observations he made of heredity, variation, and hybridism can still be accepted, though at other points we can now see that Darwin confused genetic and physiological processes.

It was the German biologist August Weismann (1834–1914) who from the late 1870s onwards did most to convince the scientific world that the hereditary substance, or 'germ plasm' as he called it, is passed from generation to generation without being influenced by bodily changes acquired through exercise, or lack of it, injury or disease. Weismann's idea of the 'continuity of the germ plasm' was based on observations of phenotypic variations of the kind Darwin made, and on an increasing microscopic knowledge of the structure and function of cells, and of cellular contents such as the nucleus. Most scientists have accepted that the discoveries of Weismann and others disproved Lamarckism, but there have been some dissenting voices in the twentieth century. Two of the most famous were the Austrian geneticist, Paul Kam-

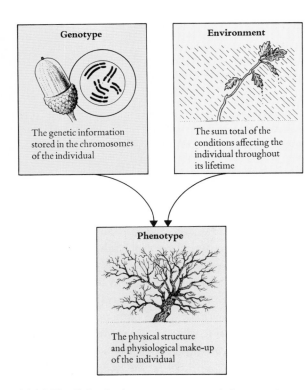

Genotype

The genetic information stored in the chromosomes of the individual

Environment

The sum total of the conditions affecting the individual throughout its lifetime

Phenotype

The physical structure and physiological make-up of the individual

(*above*) **The distinction between genotype and phenotype is fundamental to an understanding of genetic studies. The genotype includes all the information in the genes: even though some of this may be in the form of recessive genes which have no observable effects in the individual in question. The phenotype is the combined result of those genes that are expressed and all environmental influences from the moment when the egg is fertilized: it is the organism we see.**

merer (1880–1926) and the Russian biologist, Trofim Denisovich Lysenko (1898–1976).

Kammerer's work has been described by the iconoclastic Arthur Koestler, in *The Case of the Midwife Toad*. Kammerer studied acquired characters in many animals but his most famous experiment was with the midwife toad, *Alytes obstreticans*. Most toads return to water to breed and the males develop horny pads on their forefeet to enable them to grasp the female during mating in the water. The midwife toad, however, reproduces on land and the males do not have the horny pads. Kammerer claimed that when midwife toads were forced to mate in water, the males developed the pigmented 'nuptial pads' which were then inherited by the offspring. Most scientists at the time rejected Kammerer's experimental results especially when it was shown that the nuptial pads on the one specimen he exhibited were nothing more than Indian ink carefully injected under the skin. Shortly after this disclosure Kammerer committed suicide. Koestler claims that Kammerer's suicide was a direct result of the scorn heaped on him by the scientific establishment, but the great upheaval of Austrian society after the First World War may have been a contributing factor in this tragedy. It is probable that Kammerer himself did not perpetrate the fraud and it is even possible that he was 'framed'. We shall never know. In any case, Koestler insists that most of Kammerer's experimental work still stands and that it is only scientific dogmatism which rejects a Lamarckian interpretation of the phenomena which Kammerer described.

The case of Lysenko is even more striking, for he managed to gain the favour of Josef Stalin and thereby influence Russian agricultural policies for

(*left*) **A male midwife toad. Midwife toads are more adapted to life on land than other toads: intead of laying their eggs in water the male carries them twined around his hindlegs. Since mating also takes place on land, the male does not have the nuptial pads for gripping the female found in other toad species. Paul Kammerer tried to show that, if forced to mate in water, the male midwife toad developed these pads and that the pads were then inherited by the offspring.**

Jean Baptiste
de Lamarck

Charles Darwin

August Weismann

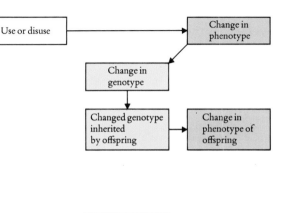

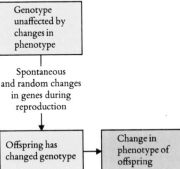

Darwin believed that the effects of habit and of use or disuse of an organ could be inherited, as did most biologists of his time. Although such ideas are often referred to as 'Lamarckism', the theory proposed by Jean Baptiste de Lamarck was rather different in that he believed in an inherent desire for improvement, *besoin*, which was the driving force of evolutionary change. August Weismann was the most influential figure in establishing the modern view that acquired characters cannot be inherited. Changes occur in the genetic material, but these are random and spontaneous and are not influenced by changes in the phenotype.

two decades from the 1930s. Lysenko was not a geneticist and Western scientists soon discredited his Lamarckian writings as fanciful. But in a Russia chronically depressed by food shortages, Lysenko's promise of quick results in breeding more productive strains of farm animals and cereal crops such as wheat and rye was seductive. The inheritance of acquired characters has the advantage that it should permit evolution to proceed far faster than the random process of variation with natural selection. Lysenko's Mendelian critics in Russia were silenced and those who refused to accept his ideas removed from positions of authority and sent to prison camps. Ironically, among those who died in the prison camps was perhaps the only man who could have produced the increase in crop productivity desired by Stalin. That man was Nikolai Ivanovich Vavilov, who in his work on the origins of crop plants laid the foundation for much of modern plant breeding. Lysenko's predictions did not work out in the end and with the demise of the Stalinist regime Lysenko himself fell from favour.

These and other twentieth-century episodes attest to our perennial fascination with Lamarckism. There seems to be something deeply rooted in the human psyche which wants the strivings and achievements of our generation to be perpetuated in the genes of those to come. But, as we shall see, what is called the Central Dogma of molecular genetics states that *somatic* (bodily) events cannot be translated into changes in the individual's genetic information. Darwin was incorrect in accepting use-and-disuse inheritance, though correct to see that the theory of natural selection did not hinge on this issue.

Darwin was also mistaken in his discussions of what is called 'blending inheritance', and various critics pointed out that blending inheritance posed problems for the concept of evolutionary change. Darwin held that each parent generally contributed equally to the offspring's physical composition, so that any inherited attribute such as height

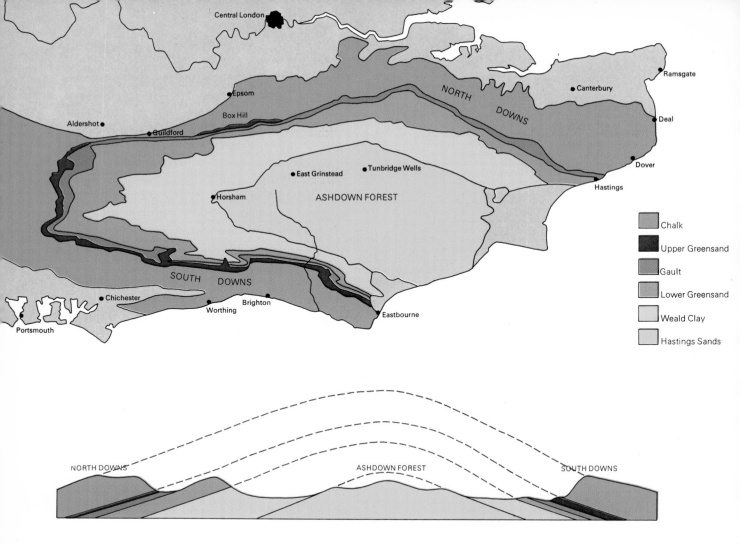

	Chalk
	Upper Greensand
	Gault
	Lower Greensand
	Weald Clay
	Hastings Sands

In the first edition of *The Origin of Species* Darwin included a calculation of the age of the Weald, the broad strip of wooded land between the North and South Downs in southern England. The Weald represents the eroded remains of a range of hills and Darwin, who wrongly believed that they had been eroded by the sea, calculated how long it had taken for them to be worn down. Basing his calculations on an estimated rate of erosion of half an inch per century he came up with the figure of 306,662,400 years.

or coloration would, in general, be the same for the offspring as for the average of the two parents. But as one of Darwin's critics named Fleeming Jenkin pointed out in 1867, this view has serious consequences for the perpetuation of evolutionary novelty. For if blending inheritance occurs, any hereditary innovation which turns up in an individual will be swamped in successive generations: its offspring will possess the novelty only half as strongly, the next generation to only one-

quarter the degree, and so forth until soon the novelty's effects are no longer expressed. Darwin believed that use-and-disuse heredity could partially solve the problem, since it heightened the rate of inherited variability. Other than this he could only insist that variation does occur – 'due to causes of which we are quite ignorant' – and that for a new variety or species to emerge the same variation must recur repeatedly and the population be isolated to prevent the variation from being swamped.

Nevertheless, blending inheritance appealed to Darwin, since it accorded with his more basic hypothesis that *Natura non facit saltum* (Nature does not make leaps). This hypothesis was also the foundation of his belief that evolutionary change always occurs gradually and slowly. His theories clearly benefited from the revolution in thinking about geological time mentioned at the beginning of this section. But Darwin's single excursion into an attempted quantitative analysis of geological time opened the door to another source of uneasiness for him. In the first edition of *The Origin of Species* he included a calculation, based on estimated rates of erosion, suggesting that the Weald, the broad strip of wooded land that lies between the North and South Downs in England and represents the eroded remains of a range of hills, must have required about 300 million years to have been worn down to its modern level. Were that true, he reasoned, the earth itself must be much older. 'How incomprehensibly vast have been the past periods of time' he wrote in the first edition of *The Origin of Species*. Even as the first edition was appearing, however, physicists such as William Thomson (later Lord Kelvin) were applying thermodynamics to the question of the earth's age. Their figure, based on rates of heat loss from the earth's surface, put the outside limit of the total age of the earth at about 200 million years: less than Darwin had calculated for the formation of the Weald. While Darwin quietly dropped this calculation from later editions of *The Origin of Species*, the problem of time continued to worry him. Various critics who did not accept evolution used the work of Thomson, the most prestigious physicist in Britain, as a weapon against Darwin, who in turn had to admit that the maximum of 200 million years since the consolidation of the earth's crust, allotted to him by Thomson, was hardly sufficient for the development of the varied forms of life.

Despite this problem, Darwin stood his ground and neither abandoned his theory nor lapsed into speculation about sudden saltatory changes which might miraculously speed up evolutionary rates. The problem was not solved until after Darwin's death, when the discovery of radioactivity pointed to another source of heat from within the earth's core, thereby demonstrating that Thomson's figures were far too conservative. Rates of radioactive decay also provide present-day geologists

with a method of dating rocks, so that the age of strata and fossils, which Darwin was unable to calculate, can now be fairly accurately determined. The neat culmination to Darwin's dilemma with time derives from the fact that one of his sons, the distinguished astronomer Sir George Darwin, was among the first to suggest that radioactivity must be taken into account when revising estimates of the earth's age. Modern estimates place the age of the earth at about 4,600 million years. Unfortunately Darwin's calculations on the Weald proved to be wrong: it is now known to be between 20 and 30 million years old.

These problems – of the fossil record; of the origin of complex adaptations and major biological groups; of heredity and variation; and of the time required for evolutionary processes to occur – are just some of the ones which Darwin faced. Many others can be gleaned from *The Origin of Species*. In successive editions of his great work Darwin strove to incorporate new knowledge and to answer his critics. Although *The Origin of Species* changed substantially between the first and the sixth editions, Darwin's primary aim remained constant: to demonstrate the power of natural selection in the gradual formation of new species. Modern science has amply demonstrated the validity of Darwin's confidence in his own discoveries.

Darwin and us

Darwin brought many different kinds of information to bear on the question of evolution, among them: heredity and variation, fossils, geological formations, geographical distribution, embryology, taxonomy and homology. Reference to a recent work on evolution (several are recommended on p. 231) will show how much this remains the same today. Since Darwin's time, there have been advances in all these individual disciplines, and some have changed beyond recognition in the past century. But it was Darwin who first showed us where to seek the evidence for evolution. In this section I shall examine a few of

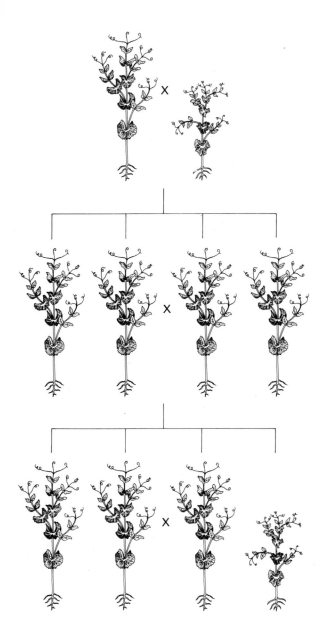

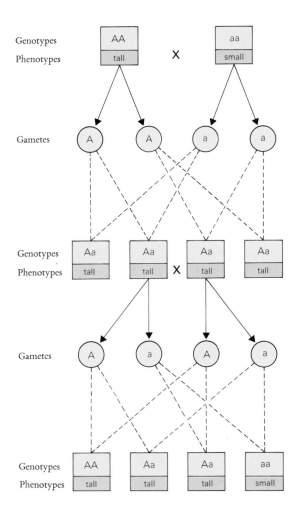

Genotypes	AA	X	aa
Phenotypes	tall		small

Gametes: A A a a

Genotypes	Aa	Aa	Aa	Aa
Phenotypes	tall	tall X tall	tall	

Gametes: A a A a

Genotypes	AA	Aa	Aa	aa
Phenotypes	tall	tall	tall	small

The diagram on the left shows one of Gregor Mendel's classic experiments with peas. If a pure-breeding tall variety was crossed with a pure-breeding short variety all offspring in the first generation were tall. When two of these plants were crossed, three tall offspring were produced for every short one. Mendel saw that these results could be explained if the characteristics were inherited as particles, of which each plant had two, and that one character was dominant to the other. In this case the tall character (A) was dominant to the short character (a). His interpretation of the results, which has been amply confirmed by subsequent research, is shown diagramatically on the right. Only a few pea plants are shown here but in fact Mendel made the cross many times and planted thousands of seeds of the successive generations to get his result, since the ratios would only be apparent when large numbers of plants were studied.

the ways in which modern research has put together some of Darwin's puzzles and answered further questions which Darwin could not even formulate.

The primary area where our understanding of evolution transcends Darwin's is in genetics. We understand the principles if not all the details of heredity, and molecular genetics is undoubtedly the single greatest achievement of the life sciences in our century. Since modern genetics derives from the work of the Austrian monk Gregor Mendel (1822–1884), Darwin could have benefited, but the importance of Mendel's observations was not realized until both he and Darwin were dead. It is a popular misconception that Mendel's 1866 paper on plant hybridism remained completely unknown until 1900. In fact there are at least eight references to Mendel's work in the scientific literature between 1866 and 1900, but it made no impact on the scientific community because Mendel's theory seemed to be at odds with general observations of inheritance. Those who had read Mendel's paper thought his instances of *particulate inheritance* were exceptions to the general rule. Only at the turn of the century was the significance of Mendel's theory recognized.

Mendel's discoveries and concepts are summarized in the diagram opposite so we need only mention here that his investigations of the inheritance patterns of garden plants, particularly peas, led him to propose the idea of particulate inheritance. Mendel noticed, when studying traits

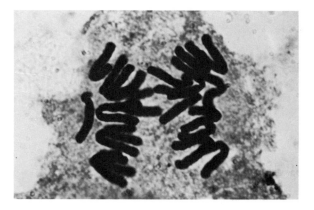

The chromosomes separate during cell division. Observation of the behaviour of the chromosomes during the cell division known as meiosis, which precedes sexual reproduction, suggested that they might be involved with the inheritance patterns that Mendel had observed.

such as flower colour, plant height, and seed shape and texture, that parental contributions are unequally expressed: blending does not occur.

To explain his results Mendel proposed that these traits were inherited as 'elements' (which we now call *genes*), one element being received from each parent. Certain elements are *dominant*, others are *recessive*, so that if the offspring has either two dominant elements, or one dominant and one recessive element, the dominant trait will be observed. Only if two recessive elements are present will the recessive trait appear.

Mendel deliberately chose simple 'either-or' characters for his experiments, i.e. characters with clearly distinguishable forms and no intermediates. Characters which are inherited in this fashion are called Mendelian. Most characters, however, are controlled by a number of genes acting together and modifying each other's activities, and so simple ratios will not be observed when their inheritance is studied, although each separate gene is inherited in a Mendelian fashion. This is why particulate inheritance was at first thought to be the exception rather than the rule.

Mendel himself developed the concepts of dominance and recessiveness, and his work embodies a clear distinction between *genotype* and *phenotype*. We observe the phenotype (like seed shape or eye colour), whereas our knowledge of the underlying genotype has had to be arrived at by subtler means.

Gregor Mendel, the Austrian monk who discovered particulate inheritance patterns in plants, notably peas. Strangely enough a statistical analysis of Mendel's results made by Ronald Fisher in the 1930s showed that Mendel's figures in his later experiments were, in statistical terms, too good to be true. The implication is that once Mendel had formulated his theory those experiments which failed to give the expected ratios were ignored, either by Mendel or his gardeners.

The behaviour of the chromosomes in sexual reproduction. This diagram does not show the processes in full but highlights those events relevant to Darwin's observations on inheritance.

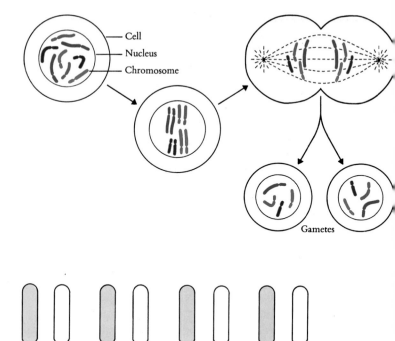

1. Every cell in the body has its own set of chromosomes. The number of chromosomes varies from species to species but is always even because each chromosome has a *sister chromosome* with which it can pair. Of this pair of chromosomes one was derived from the mother and one from the father. The genes are carried in a line along the chromosomes. Some genes are *dominant* (A) and others *recessive* (a); the latter will only be expressed if paired with an identical recessive gene. Before sexual reproduction special cells known as *gametes,* with only half the usual number of chromosomes, are formed by *meiosis,* or reduction division. Later two gametes, an egg from the female and a sperm from the male, unite to give a cell with a full set of chromosomes which then develops into an embryo.

2. During meiosis the chromosomes pair and lie alongside each other before being separated into different cells. While paired, the chromosomes may break and rejoin with their sister chromosomes. This is known as *crossing over* and results in a reassortment of genetic material; it is an important source of variation in the offspring. The closer together two genes are on a chromosome the less likely it is that they will become separated by crossing over. This means that they will almost always be inherited together, a phenomenon known as *linkage.*

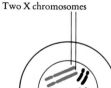

Female

Two X chromosomes

Male

X chromosome

Y chromosome

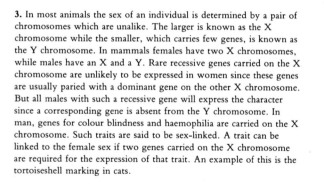

3. In most animals the sex of an individual is determined by a pair of chromosomes which are unlike. The larger is known as the X chromosome while the smaller, which carries few genes, is known as the Y chromosome. In mammals females have two X chromosomes, while males have an X and a Y. Rare recessive genes carried on the X chromosome are unlikely to be expressed in women since these genes are usually paried with a dominant gene on the other X chromosome. But all males with such a recessive gene will express the character since a corresponding gene is absent from the Y chromosome. In man, genes for colour blindness and haemophilia are carried on the X chromosome. Such traits are said to be sex-linked. A trait can be linked to the female sex if two genes carried on the X chromosome are required for the expression of that trait. An example of this is the tortoiseshell marking in cats.

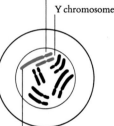

Site of sex-linked gene: no corresponding gene on Y chromosome

By the time Mendel's work was properly appreciated in 1900, microscopists had a good idea of the events taking place in two kinds of cellular division, *mitosis* and *meiosis*. In mitosis certain strands in the cell nucleus, called *chromosomes* can be seen to divide, and each daughter cell receives a full complement. Mitosis is the ordinary type of cell division. But the sex cells undergo a more complicated process of division, called meiosis. Meiosis leads to the formation of *gametes*, that is eggs or sperm. It involves two separate divisions of the nucleus but only one duplication of chromosomes, the result being four daughter nuclei, each possessing only half as many chromosomes as the parent cell. During reproduction each parent contributes one gamete, so that when the egg and sperm unite the product is a fertilized egg with a full set of chromosomes.

Observation of the way the chromosomes behave during meiosis led to the realization that they might well be involved in the patterns of inheritance Mendel had studied. In 1909 the Danish biologist Wilhelm Johannsen called the individual Mendelian unit of heredity a *gene*, and experimental work by the American geneticist Thomas Hunt Morgan and others established, at about the same time, that genes are located on the chromosomes. Morgan's experiments were performed principally on the fruit-fly *Drosophila melanogaster*, and fruit-fly experiments continue to provide much new genetic information.

Each species of plant or animal has a characteristic number of chromosomes in its cells. *D. melanogaster* has eight; human beings have forty-six. However it is more accurate to say that human cells possess twenty-three pairs, for each chromosome has its sister chromosome, opposite which it lines up during cell division. The nucleus of each somatic cell thus contains two complementary sets of genes, whereas a gamete contains only a single set, i.e. one member of each pair of chromosomes. The significance of this in relation to Mendel's theories is obvious. Each member of a pair of homologous chromosomes carries a string of genes coding for the same characters. Just as each chromosome has a sister chromosome, each gene has what might be called a 'sister gene'. These two genes are not necessarily identical: they control the same character but they may produce entirely different effects on that character. Of each pair of genes, one came from each parent. An offspring may inherit an identical gene from each parent; it is then said to be *homozygous* for that gene. If each parent contributes a different corresponding gene, the offspring is *heterozygous*. These terms refer to the genotypes only, for dominance masks the phenotypic expression of heterozygosity, though the condition may reveal itself in the appearance of recessive traits in the offspring.

These concepts are also important in understanding sex, for the sex of an individual is genetically determined at fertilization. Men and women have twenty-two pairs of chromosomes in common, but the chromosomes of the twenty-third pair – called the sex chromosomes – are different for the sexes. Women have two X chromosomes (XX), whereas men have one X and one Y chromosome (XY). Human egg cells (with twenty-three chromosomes) all contain an X chromosome, whereas sperm cells contain either an X or a Y chromosome. The sex of the offspring is determined by the kind of sperm (X or Y) that fertilizes the egg.

Y chromosomes are much smaller than X chromosomes and carry fewer genes. Because women have two X chromosomes, they rarely suffer from recessive genetic disorders carried on the X chromosome. They can carry a recessive detrimental gene on one of their X chromosomes without suffering any personal consequences. However, half of their sons (on average) will suffer from the expression of the gene, for a son's Y chromosome will not carry a gene complementary to the recessive deleterious gene on the X chromosome. Traits inherited in this way are said to be *sex-linked*. Haemophilia is the best-known sex-linked disorder; it was known in Darwin's day, but its

Tortoiseshell cats are always female. The pattern is produced when a gene for black or tabby fur and a gene for ginger fur are both present. (There is no dominance between these two genes, but instead of combination of their effects.) Since these genes are carried on the X chromosome a male, which has only one X chromosome, cannot have a tortoiseshell colouring.

sex-linked pattern of inheritance could not be explained by blending inheritance.

Genetic variation occurs through the random reshuffling of chromosomes, combined with the exchange of corresponding portions of chromosome pairs during meiosis through a process called *crossing over*, (shown diagrammatically on p. 24). Crossing over considerably increases variation in the offspring. The end result of these processes is that the offspring receives some combination of the genes possessed by its parents. Occasionally, however, a *mutation* can occur and an entirely novel gene is produced. Ultimately, mutations are the fuel of evolution, for it is through them that totally new genes and hence new variations occur. Gene mutations are random events, though their

rate can be increased by exposure to radiation and certain chemicals.

Most mutations are deleterious: they lead to diminished fitness among the offspring and are therefore weeded out by natural selection. Sometimes, however, a beneficial mutation takes place and individuals carrying this mutation acquire enhanced fitness (examples of which we shall discuss below). New species are formed by the gradual accumulation of changed genotypes through the combination of genetic variation and natural selection: through 'chance and necessity', as the French microbiologist Jacques Monod has summed it up. This is a modern statement of the core of Darwin's theory. But we can be even more specific, for during the past twenty-five years, Mendelian genetics has been transformed by molecular genetics. We now understand the molecular basis of many of the biological processes described above.

In 1953 James Watson and Francis Crick announced that they had elucidated the chemical structure of the genetic material, deoxyribose nucleic acid (DNA). DNA is a long molecule consisting of a double backbone of alternating sugar and phosphate units. Attached to each sugar unit is one of four bases: adenine, thymine, guanine, or cytosine (A, T, G, C). The long double-stranded

The structure of the genetic material, deoxyribonucleic acid, or DNA, was elucidated by James Watson and Francis Crick in 1953. The DNA molecule consists of a double backbone of alternating sugar and phosphate units *(above left)*. To each sugar unit is attached a base: adenine, thymine, guanine, or cytosine. The bases pair with those on the opposite backbone; adenine always pairs with thymine and guanine with cytosine. The long molecule is twisted to form a double helix *(above right)*. When the chromosomes replicate during cell division the molecule 'unzips' and new bases are added on by pairing, so that two new molecules are formed identical to the original one *(below)*; this is how genes are passed on to successive generations. The genetic message of the DNA is contained in the sequence of the bases: three bases taken together code for an amino acid. Thus the string of bases is translated into a string of amino acids, that is, a protein. Occasionally there is a mistake during the replication of DNA and one base is substituted for another, a base is lost, or one is added. These mistakes are known as mutations.

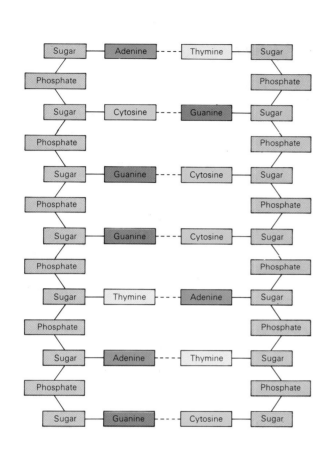

Sugar	Adenine - - - - Thymine	Sugar
Phosphate		Phosphate
Sugar	Cytosine - - - - Guanine	Sugar
Phosphate		Phosphate
Sugar	Guanine - - - - Cytosine	Sugar
Phosphate		Phosphate
Sugar	Guanine - - - - Cytosine	Sugar
Phosphate		Phosphate
Sugar	Thymine - - - - Adenine	Sugar
Phosphate		Phosphate
Sugar	Adenine - - - - Thymine	Sugar
Phosphate		Phosphate
Sugar	Guanine - - - - Cytosine	Sugar

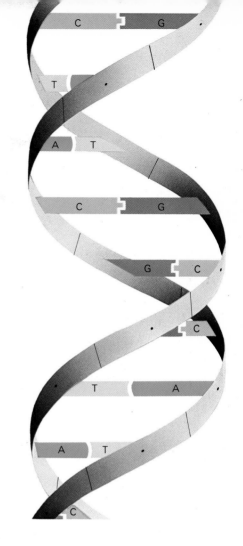

DNA molecule is twisted into a helix and the double-helical structure is such that the molecule can 'unzip' and each half-strand of the DNA can synthesize the complementary half-strand. This is what happens when the chromosomes replicate themselves during cell division. It is the sequence of the bases which spells out the genetic message of the DNA molecule. The individual 'letters' making up the messages are formed by three consecutive bases. Thus the base sequence ATTCGACGG would read ATT, CGA, CGG.

It has been determined that three particular triplet sequences mean 'stop': a gene begins and ends with a stop sign. All of the other triplet sequences code for amino acids, chemicals which are building blocks for proteins. There are twenty or so different types of amino acids, and a protein is formed of a chain of these. The particular amino acids present, and their sequence in the chain, are what give the protein its biochemical properties. Thus the genetic code is translated into a protein with specific functions in the cells of the living organism. The remarkable thing is that this genetic code appears to be universal; as far as is known, the same base triplets code for the same amino acids in all forms of life on earth. This fact in itself is strong evidence for the unity of life through common evolutionary pathways.

The other staggering implication of this arrangement is that the function of genes is simply to code for protein assembly. Properly speaking, there are no genes for eye colour or seed shape; these are merely phenotypic consequences of the protein coded for by the particular gene. The proteins themselves generally act as enzymes: substances which control chemical reactions in the body. Although the rates at which genes are translated into proteins may be controlled by a number of factors, *genetic information itself flows in only one direction: from DNA outwards.* This statement is called the Central Dogma of molecular genetics. It has been elaborated from a vast array of experimental data and seems unlikely ever to be seriously

Genes code only for proteins; all genetic effects on the phenotype are simply consequences of the action of these proteins, many of which are enzymes. The colouring of the Siamese cat is the result of a mutation affecting one enzyme involved in the synthesis of black pigment. This mutation has made the enzyme sensitive to heat, so that it is active only in the cooler parts of the body: the face and ears, the tail and the paws. These parts are dark while in the warmer centre of the body the enzyme is inactive, no black pigment is produced and the fur is light in colour.

challenged. It means that acquired characteristics are not inherited, for there is no pathway for information of the appropriate kind to reach the DNA molecules and change gene structures. It can safely be said that Lamarkism – as we defined it above – is dead, and will never be resurrected.

I hope that these brief details of one of the most fascinating stories to come out of modern biology will enable the reader to understand mutations better. Mutations generally occur when loss or addition of a base, or substitution of one base for another, takes place during replication of the DNA molecule. Additionally, a whole segment of DNA may be lost or become inverted. It is easy to see why mutations are usually deleterious, for the interactions of enzymes in the body are so finely balanced that any change is unlikely to be beneficial. Occasionally, though, mutations – with their new enzymes – can be of use to the organism. One recent example of particular importance to man is the development of penicillin-resistant bacteria.

When penicillin was first developed during World War II, it was remarkably successful in treating a wide variety of bacterial infections. It still is a very useful antibiotic, but over the past decade or two a number of kinds of bacteria have begun to produce an enzyme which breaks down the penicillin and therefore renders the bacterium relatively resistant to this particular antibiotic. Since bacteria can reproduce every twenty to thirty minutes, it will not take long for a mutant form to replace the original form, if the environment is held constant. Thus in hospitals, where the bacteria's environment is likely to contain antibiotics, resistant forms are common. In other environments where the bacteria are less likely to come into contact with antibiotics, resistance is rarer. It is not simply the mutation, but the relationship between the mutation, with its phenotypic consequences, and the environment which determines whether the fitness of the mutant organism is decreased, enhanced, or unchanged.

We have a fairly good idea of the molecular events in this example of natural selection in bacteria. Molecular genetics also enables scientists to estimate rates of evolutionary change. Haemoglobin – the protein which carries oxygen in the blood – has been most thoroughly studied from this point of view, as the structure of the haemoglobin molecule is known for many vertebrate species. Haemoglobin varies from species to species, and the comparison of the different amino-acid sequences gives one estimate of the genetic closeness of two species. As might be expected, human haemoglobin is much closer to chimpanzee haemoglobin than it is to that of a fish or reptile. By careful comparison of a number of haemoglobin structures of different species it has been estimated that haemoglobin has changed at a rate of approximately one substitution per amino-acid site per 1,000 million years, and that this rate is fairly constant. So we can count the number of changes which have occurred between different species and obtain an estimate of the affinities between them which can be usefully compared with the evolutionary trees derived from anatomical comparisons and other evidence.

The potential importance of only a small mutation at a single site on a DNA strand (called a *point mutation*) is illustrated by the human haemoglobin-S, the haemoglobin of sickle-cell anaemia, an incapacitating and frequently lethal blood disorder. Haemoglobin-S is coded for by a recessive gene, represented by the symbol Hb^S. Since the gene is recessive, sickle-cell anaemia is only apparent in individuals homozygous for the gene ($Hb^S Hb^S$). It is known as sickle-cell anaemia because the red blood cells collapse when the oxygen concentration is lowered, and they then appear sickle-shaped under the microscope. The heterozygote ($Hb^A Hb^S$) does not suffer from anaemia, but does show traces of sickling of his blood cells if the oxygen concentration is very low. The difference between normal and sickle-cell haemoglobin is only one amino acid, the result of a point mutation in the gene which codes for haemoglobin. Yet the consequences for the homozygote ($Hb^S Hb^S$) can be lethal.

Small genetic changes can thus produce lethal effects. But the question arises, why is this trait so common in some areas of Africa? Some populations in West Africa are as much as forty per cent heterozygous ($Hb^A Hb^S$) for the abnormal haemoglobin. If natural selection operates with the scrutiny which Darwin affirmed, we might expect so deleterious a gene to have been weeded out. The key lies with malaria, a serious parasitic disease spread by mosquitoes which breed in all the areas where the haemoglobin-S gene is common: in parts of West Africa, for example, malaria is almost universal. The malarial parasite enters the human blood stream when the mosquito bites, reproduces rapidly in normal individuals, and causes the classic symptoms of malaria, which may be fatal. But in individuals heterozygous for the sickle-cell trait the parasite is unable to reproduce as successfully, and the symptoms are far less severe. It is often stated that the heterozygote is resistant to malaria because the red blood cells collapse when invaded by the parasite, but things are not this simple. The real explanation may be that the mutant form of haemoglobin cannot be digested by the parasite. Therefore the heterozygotes ($Hb^A Hb^S$) do not develop malaria.

Again we see how fitness is related to the environment, for in malarial regions the heterozygote is at an advantage over both the sickle-cell homozygote ($Hb^S Hb^S$) and the normal homozygote ($Hb^A Hb^A$). In non-malarial regions the Hb^S gene would be selected against because it would confer no advantages on the heterozygote and would be deleterious when it occurred in the homozygous condition. This has happened among the Afro-American population of North America. Most of the slaves transported to North America came from the West coast of Africa and would have had a high frequency of haemoglobin-S. Since North America was never a major malarial area the Hb^S gene has now declined in frequency, but about five per cent of Afro-Americans still show the sickling trait inherited from their ancestors.

The increased resistance to malaria by individuals possessing one gene for haemoglobin-S has led to two distinct groups ($Hb^A Hb^A$ and $Hb^A Hb^S$) in some human populations. These distinct groups within a species are said to constitute a *polymorphism*. The common human blood groups (A, AB, B, O) represent another example of polymorphism in man. A more obvious instance of polymorphism occurs in the peppered moth, *Biston betularia*, found in the British Isles. In the industrial regions of England there has been an increase in the frequency of a black, or melanic, mutant form of this moth. The black mutant was very rare when it was first observed in 1849, but by

The peppered moth is a striking example of evolution in action. The original silvery form is well camouflaged on lichen-covered tree trunks but where air pollution has killed the lichen and blackened the tree trunks with soot, a black variety of the peppered moth has become increasingly common and almost replaced the original form in some places. The black mutant had begun to increase in the latter half of the nineteenth century, but, sadly for Darwin, no one knew of it at the time. This was just the evidence he needed to show the effectiveness of natural selection. Many other insect species have now been found to have black varieties in areas with air pollution; the phenomenon is known as industrial melanism.

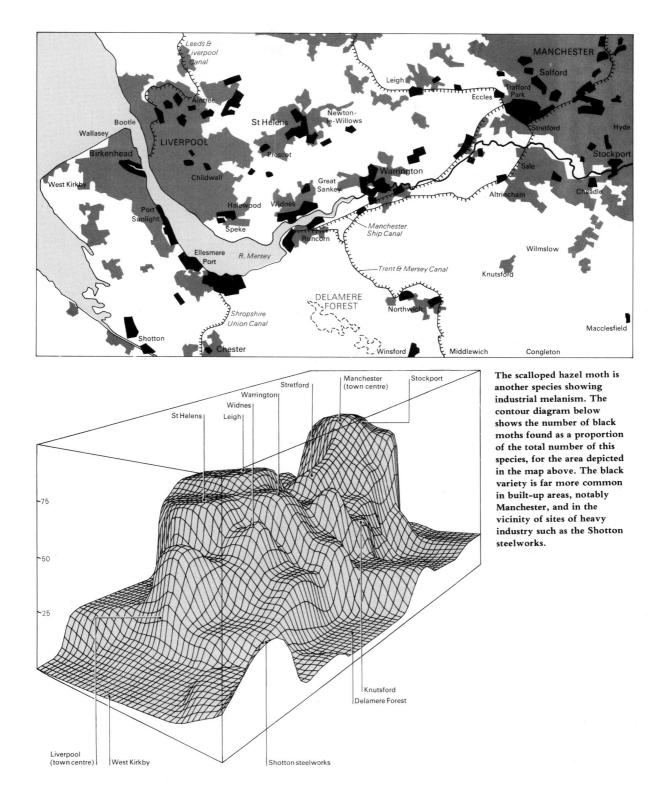

The scalloped hazel moth is another species showing industrial melanism. The contour diagram below shows the number of black moths found as a proportion of the total number of this species, for the area depicted in the map above. The black variety is far more common in built-up areas, notably Manchester, and in the vicinity of sites of heavy industry such as the Shotton steelworks.

the end of the nineteenth century it had become the most common form in the heavily industrialized area around Manchester and Liverpool. The ordinary form is silvery-white, mottled with black, and had formerly been well camouflaged from its bird predators on the lichen-encrusted tree trunks where it rested during the day. With industrial pollution of the air the lichen was killed and the tree trunks blackened by soot, with the result that the melanic form was better camouflaged than the pale form. Heavier predation of the pale form than of the melanic resulted in a gradual increase in the mutant and a decrease in the original type. This response to pollution has since been observed in a number of different species of moths, in some other insects, and in certain spiders. Industrial melanism, as it is called, has also been found in parts of North America and in continental Europe, particularly the Ruhr valley in Germany. In Britain, with the introduction of anti-pollution laws, the melanic forms have become rather less common than they were, and the light forms have begun to make a comeback in some areas.

The rise of the melanic form of the peppered moth is one of the best examples we have of evolution occurring by variation combined with natural selection in a changed environment. Ironically this instance of natural selection in action – just the evidence which Darwin needed – was going on while *The Origin of Species* went through its several editions, but no one was aware of it at the time.

The formation of a polymorphism within a particular species falls far short of *speciation* (species formation), though Darwin always insisted that a permanent variety should be viewed as a species-in-the-making. Modern genetics has given us a much better idea of speciation than Darwin had, though he perceptively stressed the crucial importance of *isolation* in the process. Geographical isolation as an evolutionary force was impressed upon Darwin through his experiences on the Galapagos Islands, which he visited during the voyage of the *Beagle*. Although only about five hundred miles from the west coast of South America, the islands have a unique flora and fauna. Furthermore some species vary markedly from island to island. The governor of the Galapagos Islands could identify the island of origin of any of the giant tortoises found there simply from the shapes and markings of their shells. The many different species of finches varied most strikingly in the size and shape of their beaks, a characteristic obviously related to the way in which the particular species of finch obtained its food. Since the Galapagos Islands are volcanic and of geologically recent origin, their isolation from the South American mainland and (through deep waters and swift currents) from each other demonstrated powerfully to Darwin how a new community could develop. Geographical isolation not only prevents the dilution of mutant genes which arise within the isolated group, it can also expose the group to new environmental conditions, where different variations are favoured.

Of all the places H.M.S. *Beagle* visited during her five-year voyage, it was the Galapagos Islands that most impressed Charles Darwin. The animals found there were unique, but quite clearly related to those of the South American mainland. There were also slight differences between animals of the same species on the different islands. The governor of the Galapagos Islands told Darwin that he could identify the island of origin of any of the giant tortoises by the shape and markings of its shell.

Mating between individuals of different species is a rare event in the wild, but it can happen, as when this male pine hawk moth and female lime hawk moth mated. The lime hawk moth later laid eggs that failed to hatch, so that her reproductive effort was wasted. Normally differences in courtship behaviour would prevent such interspecific mating. It is obvious that natural slection would favour isolating mechanisms, such as courtship differences, which prevented interspecific mating, and ensured that reproductive effort was not wasted in this way.

Many biologists today believe that new species almost always develop from a small population which for some reason has become isolated from the parent group. Ernst Mayr has been particularly instrumental in establishing this theory of speciation, which is known as *allopatric speciation* meaning, loosely, 'in another place'. Small, isolated groups encounter new ecological pressures and favourable mutations can spread quickly. Mayr believes that, under propitious circumstances, species formation can be a relatively rapid process. There is much field evidence supporting the importance of allopatric speciation, and the speed of the process may help to explain why transitional fossil forms are so rare.

There are other mechanisms by which reproductive isolation can occur, sometimes without the necessity of physical separation of the two groups. The mechanisms can involve seasonal variation (closely allied species may breed at different times of the year); behavioural complexities (elaborate courtship rituals are one important means whereby members of a species recognize their own); mechanical factors (the reproductive organs of closely allied species may be so structured as to prevent copulation); ecological preferences (closely related species may prefer different environments, such as, with leaf-eating insects, the leaves of different species of plants, and so rarely come into contact even though inhabiting the same region); or physiological incompatibilities (the offspring of interspecific crosses may be weak or the embryo may fail to develop beyond a certain stage; alternatively, inter-species hybrids may be healthy but sterile, or of reduced fertility, so that they are unable to reproduce).

Darwin recognized these various mechanisms but did not believe they could be favoured by natural selection. He ascribed all instances of reproductive isolation to differences between the species which had arisen incidentally as they diverged. He was partly correct: the sterility of hybrids, such as the mule, is caused by a divergence in structure and number between the chromosomes of its parents, the horse and the ass. The mule itself is very vigorous, but because the chromosomes do not pair up in an orderly fashion meiosis cannot proceed regularly. The gametes produced have too many or too few chromosomes and so the mule is sterile.

It is now realized that, in such a situation, natural selection could favour an isolating mechanism which acted earlier on in the reproductive process to prevent the sterile hybrid from being formed. Suppose that two species that could interbreed but produced hybrids that were sterile or of reduced fertility coexisted in the wild: every time a male and female of the different species mated their reproductive effort would be wasted, in evolutionary terms, for the hybrid, being infertile, would propagate their genes no further. If a mutant individual appeared which was more discriminating and only ever mated with those of its own species, it would, on the whole, tend to pass more of its genes, including the 'discriminator' gene, on to subsequent generations because all its offspring would be fertile. Natural selection would favour the discriminating mutant, and the mutation would, in time, spread through the population.

Speciation is a genetic phenomenon, and we have seen a few of the ways in which our knowledge of genetics – Mendelian and molecular – has enhanced our appreciation of the view of life which Darwin first elaborated. But where did life begin? For Darwin there were almost no fossils to be found before the Cambrian. (He believed that numerous fossils of a tiny creature, which was given the name *Eozoon*, were present in certain Precambrian strata in Canada, but these were later shown to be nothing more than crystalline formations.) Yet he affirmed his belief in the origin of life from either a single source or a few simple forms. Both the logic of his theory and his own deeply felt convictions required the existence of life long before the Cambrian, with its abundant and varied fauna, which includes representatives of nearly all the present-day invertebrate phyla.

Until very recently, the apparently sudden advent of fossils in the Cambrian was a source of some concern for palaeontologists. Now, however, careful micropalaeontological studies have revealed numerous single-celled fossils in very ancient Precambrian rocks, and in the Upper Precambrian the fossils of some simple multicellular organisms, such as worms and jellyfish, have been found. (It still remains to be explained why single-celled organisms which had then existed for between 2,000 and 3,000 million years, gave rise in the space of a mere 200 million years to the many different and complex creatures of the Cambrian, with the major burst of activity compressed into the last 50 million years of this period. The so-called 'Cambrian explosion' is a problem

to which we shall return.) This research has extended the record of life on earth from about 570 million years (the age of the earliest Cambrian deposits) to about 3,350 million years, and has yielded much new insight into the nature of the earliest living things.

These earliest cells had no nuclei, and scientists call such a cell a *prokaryote*. There are two groups of prokaryotes: bacteria and blue-green algae. All higher plants and animals, fungi, protozoa and algae other than blue-green algae are *eukaryotes*. Within each cell they have a membrane-bound *nucleus* which contains the chromosomes, and various other structures which prokaryotes lack. Whereas prokaryotes may occasionally acquire small portions of genetic material from each other, eukaryotes have an efficient and regular process for genetic exchange: sexual reproduction. The evolution of eukaryotes from prokaryotes was one of the most significant events in the earth's history.

Life undoubtedly originated in an oxygenless environment, and certain of the earliest organisms, the blue-green algae, played a crucial part in producing our modern atmosphere, since they release gaseous oxygen, by the process of photosynthesis. It took about a thousand million years for these microscopic organisms to transform the atmosphere into an oxygen-rich one such as we now breathe. The atmospheric changes which these first organisms effected illustrate an important biological generalization: plants and animals do not simply exist in environments, they also change them. The earth is not simply the stage on which evolution is performed: it is also an actor in the drama.

Much of our knowledge of the earliest evolutionary events must be obtained through careful inference. But a famous series of experiments by Stanley L. Miller and Harold C. Urey, simulating what might have been the condition of the primeval planet, has directly demonstrated that many of the organic building blocks of life can be produced from a mixture of water, methane, ammonia, and hydrogen. Electrical discharges (as could be initiated by lightning) passed through such a mixture, sealed off from the atmosphere, yield a number of more complicated organic substances, such as aldehydes, carboxylic acids, and the basic components of protein: amino acids. It is of course a long distance from this 'organic soup' to a living cell, and no one has yet succeeded in creating life in the laboratory. But the Miller-Urey experiments and much basic biochemical research over the past two decades have elucidated some of the probable steps in early biochemical evolution.

It is important to realize that such syntheses of organic compounds were only possible in an oxygenless atmosphere. Furthermore ultraviolet radiation, as well as lightning, was probably an important source of energy in the syntheses which took place. As oxygen (O_2) built up in the atmosphere a layer of ozone (O_3) developed above the oxygen, and this now screens out most of the ultraviolet radiation from the sun. For these reasons, among others, a living organism could not arise *de novo* on earth today.

But in Darwin's time it was widely believed that such 'spontaneous generation' could occur, since tiny creatures could be observed under the microscope, in liquid in which vegetable matter had been infused. In fact Lamarck and other early evolutionists such as Robert Chambers relied on spontaneous generation to maintain the supply of lower organisms. They saw evolutionary progress as inevitable: an inexorable force driving every creature towards higher and higher levels of perfection, and for this reason their theories could not comprehend the continued existence of 'lowly forms'. At the very time *The Origin of Species* was published Louis Pasteur was conducting his experiments demonstrating that spontaneous generation did not occur, but his findings presented no problems for Darwin, who did not see why a lower organism should not persist unchanged if well adapted.

Louis Pasteur in his laboratory. Pasteur's demonstration that spontaneous generation of life does not occur was made at the time *The Origin of Species* appeared, but Darwin still kept an open mind on the subject as late as 1872. Early evolutionists such as Lamarck believed biological evolution to be inevitable and depended on spontaneous generation to maintain the supply of lower organisms. Unlike them, Darwin saw no reason why a simple organism should not persist unchanged if well adapted to its environment.

To return to the problem of the 'Cambrian explosion': various theories have been put forward in an attempt to explain this phenomenon. Some argue that the 'explosion' is illusory, and that the early radiation of the invertebrates took place during the Precambrian but in a restricted locality, or one where fossilization was unlikely. This was then followed by a relatively rapid migration owing to some change in conditions. Another

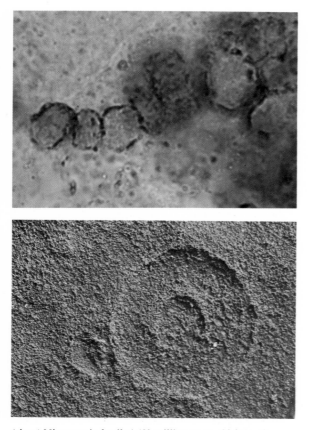

(above) **Microscopic fossils 1,600 million years old from the Amelia Dolomite rocks of Australia. Until fairly recently no fossils were known in strata older than the Cambrian. Now, careful search has revealed microfossils such as these in Precambrian rocks, and has extended the record of life on earth by nearly 3,000 million years.**

(below) **In the Upper Precambrian a few fossils of simple multicellular animals, such as jellyfish, have been found. Such animals must have been far more common than the sparse fossil record suggests, for soft-bodied organisms are only fossilized under exceptional circumstances.**

was triggered by the evolution of eukaryotic cells, which, with their chromosomes contained within a nucleus, carry out an efficient form of sexual reproduction, unknown among prokaryotes. This refinement of sexual reproduction allowed a vast increase in the rate of genetic variation which resulted in a burst of adaptive radiation.

But when did eukaryotes first develop? Some palaeontologists claim it was as long ago as a thousand million years before the beginning of the Cambrian, in which case the above theory is not very convincing. But there is disagreement over the interpretation of these ancient microfossils, and other palaeontologists think that eukaryotes appeared much later in the Precambrian.

Given the fact that scientists are trying to understand events which took place thousands of millions of years ago, we shall never know the complete story of the origin of life. But evolutionary theory is probably the most firmly established theory in the life sciences today. Tomorrow's observation and research may yield the unexpected. Who could have predicted the existence of a living coelacanth, a kind of fish that was thought to have been extinct for some 100 million years? On the other hand the range and variety of phenomena explicable by the theory of evolution is staggering. Darwin initiated a multidisciplinary research programme which has, during the past century, answered many of the questions raised by *The Origin of Species*. But contemporary evolutionary biology is not without its own unanswered queries and areas of controversy; to some of these we can now turn.

Research and controversy: evolutionary biology today

In sketching a century of Darwinian biology, I may have given the impression that most of the major evolutionary issues have been tidily resolved. This is not true, for every branch of science is constantly changing, and controversy lends force and excitement to the scientific enterprise.

group of theories invokes a change in the environment as a trigger for a rapid radiation of forms. The most plausible of these argues that oxygen concentration only reached a level high enough to support active multicellular animals in the late Precambrian. A third group of theories attempts to explain the 'explosion' simply in biological or ecological terms without invoking any climatic change. One such theory claims that this radiation of forms

However, we have already seen several points which do seem to be established beyond doubt: the fact of evolution; DNA as the genetic substance; the Central Dogma of molecular genetics. Within each of these areas of knowledge there are, of course, still unanswered questions, but each area contains a hard core of fact in which we can place confidence.

Today there is research of evolutionary import being carried out in every single discipline which Darwin combed for evidence bearing on his theory. Anatomy, embryology, physiology, geology, palaeontology, genetics, taxonomy, and ecology all have something to contribute to evolutionary theory. To take but a few examples, continental drift theory states that the present-day continental land masses were at one point connected to each other. There is much evidence to suggest that this was so and that the continents are still drifting apart very slowly. What is currently known about continental drift is summarized in a series of maps on pp. 188–9. This theory has important consequences for observations which nineteenth-century geologists made about the similarity of fossil species found in ancient strata in the Old and New Worlds, particularly in the Southern Hemisphere. Creationists would easily explain this as a result of worldwide creation, but it is now possible to understand these phenomena in terms of continental drift. Likewise, new insights into the taxonomic relationships of species can be gleaned by using computers to handle the large quantities of detailed information being compared. The science of ecology, which Darwin was the first to grasp in all its complexity, has acquired new urgency as man changes the environment at an increasing rate. Molecular genetics is presently one of the most exciting areas of biological research, and there is much to be learned about gene regulation, gene expression, and the details of genetic structure and function in more complicated organisms. So basic is the theory of evolution to the life sciences that there is little biological research that is not, at some level, relevant to the subject. The more we learn about organisms, the more we can understand their evolution.

More specifically, though, there are four or five major areas of controversy in modern evolutionary biology which should be mentioned. These include the problem of neutral mutations; the universality of natural selection as the evolutionary mechanism; the evolution of sex; the origin of altruism; and 'sociobiology', or the possibility of a synthesis of evolutionary biology and the social sciences. We shall briefly consider each of these issues.

We have already seen that a mutation involves some change in the DNA molecule, which in turn affects the proteins whose synthesis the DNA regulates. Mutations lead to altered proteins and, in Darwinian natural-selection terms, beneficial mutations are favoured, deleterious mutations weeded out. But what happens to a mutation without selective consequences? What if the new protein molecule with its one substituted amino acid still performs its function with the same efficiency? Under ordinary circumstances, with a large, freely-breeding population, such a mutation will not increase in frequency. Some biologists, however, hold that quite a large percentage of mutations are neutral, in which case this represents an important non-Darwinian source of evolutionary change, (i.e. a source of change unaffected by natural selection), even though neutral mutations would not in themselves lead to new adaptations.

The differences among proteins due to these mutations can be detected by a range of biochemical techniques. Over the last ten years or so, geneticists using these techniques have discovered far more protein variation in populations than was ever thought possible. The existence of more than one protein form in a population is called a *protein polymorphism*. For such a polymorphism to be maintained by natural selection a proportion of the population must die, as in the sickle-cell-anaemia/malaria example. If all the protein polymorphisms which have been found were being maintained by natural selection, then calculations show that the number of individuals which would have to die is often greater than the reproductive capacity of the organism.

Evolutionists are thus forced back to the conclusion that these protein variants are neutral. If this is so they should not change in frequency from one generation to another, and all proteins should show about the same amount of polymorphism. But in fact some proteins are far more variable than others: why is this? The question of neutral mutations continues to be one of the central problems in modern evolutionary theory.

Neutral mutations represent one way in which factors other than natural selection may operate on the gene pool; another way is by *genetic drift* but this only affects small populations. In the absence of natural selection a gene should not change in frequency from generation to generation. In a large population this will generally be so, but in a small population gene frequencies can change due to chance. If you toss a coin a hundred times it is highly unlikely that you would get a hundred heads and no tails, but if you toss a coin twice it is quite likely that you would get two heads and no tails. This chance effect when numbers are small is what produces genetic drift.

In the study of small populations, of man as well as other organisms, the effects of drift can often be observed. The frequency of characters in these small populations may differ greatly from the frequency in the large populations from which they have become isolated. For example, in large human populations most genetic disorders are rare but in a number of closed religious communities, such as the Amish in Pennsylvania, the same genetic disorders, such as Troyer Syndrome, a form of muscular dystrophy, are often quite common. In these cases the communities have usually been founded by small numbers of individuals some of whom would have been carrying rare recessive genes in the heterozygous condition. In

the small populations derived entirely from these founders, such genes have become common by the chance process of genetic drift, and individuals homozygous for the recessive genes, and thus displaying the recessive trait, occur far more frequently than in the outside world. Genetic disorders are only the more obvious manifestations of an overall change of gene frequencies in the

Two members of the Amish, a closed religious community in Pennsylvania. The gene frequencies among such small and genetically isolated populations often differ from those of the population at large, because genetic drift has occurred. Genetic disorders that are generally rare may be quite common within the closed community.

gene pool of the small population.

The debate about neutral mutations and genetic drift represents part of the argument between those who would view all the traits which an organism displays as having adaptive significance, and those who believe that some structures are simply fortuitous or result from developmental constraints of which we may not always be aware. An example of the latter is the human chin. This structure, which none of the great apes possesses, was a puzzle to biologists until it was realized that it had no adaptive significance at all. The chin is just the outcome of different growth rates operating in the upper and lower margins of the lower jaw. Both growth rates have slowed down relative to our ancestors, but the growth rate of the upper margin has slowed down more, producing a jutting out of the lower margin: in effect, a chin. This is an example of a general phenomenon known as *allometry*.

One example of this sort of argument about adaptive significance concerns the varied and intricate patterns seen on the shells of many molluscs, which some evolutionists believe are not adaptive. However, the functional significance of any character may not be obvious unless the behaviour and ecological relations of the organism are fully understood. The polymorphic banding patterns and coloration of a certain species of land snail, *Cepea nemoralis*, were long believed to have no adpative significance. These snails exist in pink, yellow, and brown forms with five, three, one, or no blackish bands.

By careful observation of the song thrush, one of the snail's predators, Philip Sheppard and Arthur Cain were able to show that thrushes tended to take the snails which were more conspicuous, and that in some habitats one form was better camouflaged than others. On smooth grass, for example, brown-shelled snails were taken more frequently than yellow-shelled snails, whose colour concealed them, but among dead leaves brown-shelled snails were well camouflaged. In rough pasture five-banded snails were at an advantage since they blended well with the background of plant stalks. Before it can be decided that an organism possesses a characteristic which is not adaptive its biology as a whole must be carefully studied.

On the other hand, the critics of the 'everything-is-adaptive' approach have a very valid point when they say that if one is looking for an adaptive explanation it is all too easy to find one. If every characteristic is automatically accepted as being adaptive, it is not hard in most cases to construct plausible explanations. Darwin himself recognized this pitfall when he wrote:

> If green woodpeckers alone had existed, and we did not know that there were many black and pied kinds, I dare say that we should have though that the green colour was a beautiful adaptation to conceal this tree-frequenting bird from its enemies; and consequently that it was a character of importance, and had been acquired through natural selection; as it is, the colour is probably in chief part due to sexual selection.

The debates between 'selectionists' and 'neutralists' should not be taken to means that we are back in the nineteenth-century deadlock, where many scientists who accepted evolution questioned natural selection as its principle mechanism. Since the work of the population geneticists in the 1930s and 40s, the operation of natural selection on variations and mutations has been widely recognized as *the* driving force of evolutionary change and the research of many contemporary evolutionists, such as Ernst Mayr and Theodosius Dobzhansky, has tended to confirm this. But there are still biologists who are unhappy that natural selection is such an elastic concept.

Both adaptation and natural selection, though intuitively easy to grasp, are frequently difficult to study rigorously: their investigation involves not only highly complex ecological relationships, but also the advanced mathematics of population genetics. Critics of natural selection may be correct to challenge its universality, but the significance of other mechanisms, such as neutral mutations and genetic drift, are still unknown. In the meantime, it is clear that modern evolutionary theory makes as much use of natural selection as did Darwin, though population and molecular genetics have transformed the way we understand it.

Did sex and sexual reproduction also evolve through natural selection? Undoubtedly the answer is yes, but since it is probable that sexual reproduction developed between 1,000 and 2,000 million years ago, much detail remains unclear. Generally, sexual reproduction is a characteristic separating eukaryotes from prokaryotes, although prokaryotes occasionally exchange genetic information with each other, and some eukaryotes reproduce asexually. In evolutionary terms, sexual reproduction must be extremely important, since many species devote so much energy to it. The advantage of sexual reproduction is obvious, since it enormously enhances variability. An asexual organism simply produces identical copies of itself and thus its potential for change is limited, occurring only via gene mutation. In this setting, the acquisition of two favourable mutations by a single individual requires two independent events, whereas in sexually reproducing species the chances of the genes coming together through genetic exchange are much greater.

On the other hand, a female individual of a successful species who opted for asexual reproduction would more or less double her *genetic fitness* immediately. That is to say, all her reproductive effort would be invested in reproducing her own genes, instead of half of it being invested in the genes of a mate, so that, other things being equal, she would pass on twice as many copies of her genes to the next generation. One might therefore expect female members of well-adapted species to go over to asexual reproduction or parthenogenesis (and thus to become female-only populations) quite regularly. The fact that this does not happen

Are the colourful and intricate patterns on these shells *(above)* of adaptive significance to the marine molluscs that live in them? Some evolutionary biologists would argue that they are not, and that it is a dangerous mistake automatically to assume that every characteristic of every living organism has adaptive significance. But a closer study of the ecology of these molluscs may perhaps reveal a functional role for their patterns, as it did with the land snail *Cepea nemoralis (below)*. Careful observation of one of the snail's predators, the song thrush, showed that the different varieties of this snail were adapted for camouflage against different backgrounds. The brown form is best camouflaged against dead leaves, as shown here. The yellow, unbanded form is well protected on smooth grass, while the banded snails would blend into a background of plant stalks.

attests to the ever-changing nature of the environment, which requires species constantly to adapt to slightly different circumstances; this is known as *environmental tracking*, or more colloquially as the Red Queen hypothesis, after the character in *Alice in Wonderland* who had to keep running to stay where she was. The individual who kept her offspring isolated from the gene pool by reproducing parthenogenetically would doom them to eventual extinction, since, without the variability which sexual reproduction provides, they could not keep pace with the environment.

Many invertebrate animals do reproduce parthenogenetically at some time, but such a process is usually interspersed with sexual reproduction. Aphids, for example, reproduce asexually during the summer when food is in good supply. Wingless females are produced which rapidly give birth to more wingless females, resulting in a population explosion. Males are born only in the autumn, when sexual reproduction takes place and eggs are laid which overwinter and hatch in the following spring. In certain plants sexual reproduction has become relatively unimportant, but this has happened only in plants such as brambles, *Rubus* species, and dandelions, *Taraxacum* species, which

How did sex evolve? In evolutionary terms sexual reproduction must be important since so many species devote so much energy to it. Shown here are a pair of giant Galapagos tortoises and a pair of green hairstreak butterflies. The advantage of sexual reproduction is that it enormously increases the variability of the offspring. The more variability there is available the more likely it is that some of the offspring will be able to adapt to the ever-changing environment.

Many invertebrate animals can reproduce asexually, but only do so for part of the year. Here a female aphid gives birth to live, wingless females, who are themselves able to reproduce shortly afterwards. This allows a rapid build-up of the population during the summer when food is plentiful. In the autumn males are produced and sexual reproduction takes place.

There is a limit to the number of chicks that a pair of birds can raise in one season: this limit is imposed by the amount of food they can collect. If more eggs were laid there would not be enough food to go round and most or all of the chicks would starve.

are highly successful and have effective means of spreading themselves vegetatively.

Darwin argued that natural selection acts only on the individual (although he made an exception in the case of the social insects). Today evolutionary biologists have a very similar view, but in the interim a largely erroneous theory which suggested that natural selection could operate at the species level was accepted by some biologists, and achieved popular currency. This theory was proposed to explain what appeared to be altruistic acts performed by one animal for another, or a group of others. In the popularized form of the theory it was said that these acts were 'for the good of the species'.

Some such explanations can easily be dismissed with a little thought: a female bird does not protect and feed her young 'for the good of the species' but for the propagation of her own genes. (It should be kept in mind that this does not imply her motive, but only that if she did not have a genetically determined 'maternal instinct' her genes would soon become extinct since her offspring would not

survive.) Similarly, birds do not limit the number of eggs they lay to prevent a population explosion 'for the good of the species'. If they only lay four eggs they do so because they would be unable to collect enough food for five baby birds, with the result that all or most of them would starve, and thus fail to pass on the parents' genes.

Other instances of 'altruism' may not be dismissed so easily. In baboon troops one member may risk his life in the defence of the troop. In certain bird species not all individuals mate and rear their own young: those who do not do so may help a brother to gain a mate, or, in other species, may help in rearing their younger brothers and sisters. A theory which appears to explain all these instances of behaviour in Darwinian terms is *kin selection*. This theory in its simplest form states that a brother or sister shares as many of your genes (half) as a child, so that in helping a sister or brother to survive and reproduce you are helping to perpetuate your own genes (or, to put it another way, making a gain in genetic fitness) as much as if you protect your own child. The degree of relatedness of any two individuals is calculated according to the percentage of genes held in common, and predictions are made from this about the extent to which these two individuals should show 'altruism' towards each other. The theory is appealing, but recent work implies that relatedness cannot be equated with the gain in genetic fitness to be made from an altruistic act as simply as has been suggested, and that a far more complex mathematical approach is needed, which takes account, for example, of the age and reproductive potential of the 'kin'. To give a very simple illustration, there is little gain in genetic fitness to be made by rescuing an elderly brother or sister who is unlikely to survive much longer anyway.

One of the instances of 'altruism' most difficult to explain is that of a bird who gives a warning cry on seeing the approach of a hawk. This cry enables the whole flock to fly away at once, thus confusing the hawk, and reducing the number of birds captured. But the bird who gives the cry makes itself more conspicuous and therefore more likely to be killed. Only if a large number of birds in the flock are related to the one who gives the warning cry can this behaviour be explained by kin selection.

As Darwin realized, the social insects are an exceptional case, and the sterile worker insect can truly be said to be working for the good of the colony. Natural selection works on the colony as a whole, not within it, and the colony is therefore best thought of as a *hyperindividual*, with each member analogous to cells in the body. Since the workers are irreversibly sterile, their activities can hardly be labelled 'altruistic': self-sacrifice is not possible unless there is something to be sacrificed. The real puzzle with social insects rests in the early stages of their evolution.

The glib use of the word 'altruism' in these con-texts, and the disproportionate attention which the debate receives are a clue to the real issue at stake: is altruism in human beings simply another expression of evolution at work? More generally, do human values transcend the world of natural selection, or is it possible to understand our behaviour through a judicious combination of evolutionary biology, ethology (the study of animal behaviour) and social science? Harvard biologist Edward O. Wilson is the main proponent of the view that biology and the social sciences can be amalgamated into what he calls 'sociobiology'.

Wilson argues that, since some behavioural traits in man are universal, cultural transmission may not offer the complete explanation. He speculates that some, incest avoidance for example, could have a genetic basis. Wilson's ideas have attracted violent criticisms, and there is certainly a danger in extrapolating from animal behaviour to humans, or vice versa, in the uncritical fashion which characterizes much of 'sociobiology'. It should be questioned whether we can talk about the behaviour of animals in the human terms which have been widely used by sociobiologists, such as 'faithfulness' and 'philanderer', without immediately prejudicing our conclusions in favour of the implied analogy. There are certainly many pitfalls, but also much to be considered. A more rigorous approach and rather more dispassionate discussion between representatives of the differing viewpoints are required.

In the meantime this much is certain: all aspects of modern evolutionary biology can be seen as part of a research programme inaugurated by *The Origin of Species*. It is without doubt the most important biological work ever written.

(overleaf) H.M.S. *Beagle* in the waters of Tierra del Fuego, painted by Conrad Martens.

Introduction
to the Original Work

When on board H.M.S. *Beagle*, as naturalist, I was much struck with certain facts in the distribution of the organic beings inhabiting South America, and in the geological relations of the present to the past inhabitants of that continent. These facts seemed to throw some light on the origin of species – that mystery of mysteries, as it has been called by one of our greatest philosophers. On my return home, it occurred to me, in 1837, that something might perhaps be made out on this question by patiently accumulating and reflecting on all sorts of facts which could possibly have any bearing on it. After five years' work I allowed myself to speculate on the subject, and drew up some short notes; these I enlarged in 1844 into a sketch of the conclusions which then seemed to me probable: from that period to the present day I have steadily pursued the same object.

My work is now (1859) nearly finished; but as it will take me many more years to complete it, and as my health is far from strong, I have been urged to publish this Abstract. I have more especially been induced to do this, as Mr. Wallace, who is now studying the natural history of the Malay Archipelago, has arrived at almost exactly the same general conclusions that I have on the origin of species. In 1858 he sent me a memoir on this subject, with a request that I would forward it to Sir Charles Lyell, who sent it to the Linnean Society. Sir C. Lyell thought advisable to publish, with Mr. Wallace's excellent memoir, some brief extracts from my manuscripts.

This Abstract, which I now publish, must necessarily be imperfect. I cannot here give references and authorities for my statements. I can give only the general conclusions at which I have arrived, with a few facts in illustration.

In considering the Origin of Species, it is quite conceivable that a naturalist, reflecting on the mutual affinities of organic beings, on their embryological relations, their geographical distribution, geological succession, and other such facts, might conclude that species had not been independently created, but had descended, like varieties, from other species. Nevertheless, such a conclusion would be unsatisfactory, until it could be shown how the innumerable species inhabiting this world have been modified, so as to acquire that perfection of structure and coadaptation which justly excites our admiration. Naturalists continually refer to external conditions, such as climate, food, etc., as the only possible cause of variation. In one limited sense this may be true; but it is preposterous to attribute to mere external conditions, the structure of the woodpecker, so admirably adapted to catch insects under the bark of trees. In the case of the mistletoe, which draws its nourishment from certain trees, which has seeds that must be transported by certain birds, and which has flowers with separate sexes absolutely requiring the agency of certain insects to bring pollen from one flower to the other, it is equally preposterous to account for the structure of this parasite, with its relations to several distinct organic beings, by the effects of external conditions, or of habit, or of the volition of the plant itself.

It is, therefore, of the highest importance to gain a clear insight into the means of modification and coadaptation. At the commencement of my observations it seemed to me probable that a care-

ful study of domesticated animals and of cultivated plants would offer the best chance of making out this obscure problem.

I shall devote the first chapter to Variation under Domestication. We shall thus see that a large amount of hereditary modification is at least possible; and, what is equally or more important, we shall see how great is the power of man in accumulating by his Selection successive slight variations. I will then pass on to the variability of species in a state of nature. In the next chapter the Struggle for Existence amongst all organic beings throughout the world, which inevitably follows from the high geometrical ratio of their increase, will be considered. This is the doctrine of Malthus, applied to the whole animal and vegetable kingdoms. As many more individuals of each species are born that can possibly survive; and as, consequently, there is a frequently recurring struggle for existence, it follows that any being, if it vary however slightly in any manner profitable to itself, will have a better chance of surviving, and thus be *naturally selected.*

This fundamental subject of Natural Selection will be treated at some length in the fourth chapter; we shall see how Natural Selection almost inevitably causes much Extinction of the less improved forms of life, and leads to Divergence of Character. In the next chapter, I shall discuss the

(above left) **Sir Charles Lyell (1797–1875).** Darwin read his *Principles of Geology* on board H.M.S. *Beagle,* and was greatly influenced by his ideas. On his return to England he met Lyell, and they became close friends. Initially he was opposed to Darwin's theories of biological evolution, but was gradually converted. When Darwin received Wallace's paper on natural selection it was Lyell who suggested they should publish a joint memoir.

'This is the doctrine of Malthus, applied to the whole animal and vegetable kingdoms.' **In his** *Essay on the Principle of Population* **written in 1798 Thomas Malthus** *(left)* **was concerned primarily with human populations. Malthus argued that populations tend to increase indefinitely and that they can only be kept in check by food shortages, plagues and wars. Both Darwin and Wallace grasped the principle of natural selection as a direct result of reading Malthus's essay.**

complex and little-known laws of variation. In the five succeeding chapters, the most apparent and gravest difficulties in accepting the theory will be given. I shall consider the geological succession of organic beings throughout time and their geographical distribution throughout space.

Although much remains obscure I can entertain no doubt that the view which most naturalists until recently entertained, and which I formerly entertained – namely, that each species has been independently created – is erroneous. I am fully convinced that species are not immutable; but that those belonging to what are called the same genera are lineal descendants of some other and generally extinct species, in the same manner as the acknowledged varieties of any one species are the descendants of that species. Furthermore, I am convinced that Natural Selection has been the most important, but not the exclusive, means of modification.

The European mistletoe, *Viscum album*, is a partial parasite, making its own food by photosynthesis but drawing water and mineral salts from the tree on which it grows, usually an apple or a poplar. The mistletoe must not drain the host tree of resources, since if the tree dies the mistletoe dies also. Bees and flies must be attracted to carry pollen from the male flowers to the female flowers which are borne on separate plants. Finally the berries must be eaten by birds for the seeds to be spread. The berries are particularly attractive to the mistle thrush. In Darwin's words '. . . *it is preposterous to account for the structure of this parasite, with its relations to several distinct organic beings, by the effects of external conditions, or of habit, or of the volition of the plant itself.*'

(right) The numerous adaptations of the woodpecker to feeding on insects that live in wood were, to Darwin, equally inexplicable except by his own theory. Shown here is the green woodpecker *Picus viridis*.

The woodpecker is superbly adapted for chipping away wood and picking out insects from cracks. The beak is very strong with a chisel-like tip, and the neck muscles highly developed for hammering. Even more remarkable are the flexible, elongated hyoid bone which enables the tongue to be greatly protruded, and the complex muscles which push the tongue in and out, stiffen it for spearing insects or move it from side to side to explore the crevices of a rotting log.

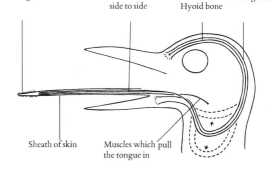

Horny barbed tip to spear larger insects

Muscles which can stiffen the tongue or move it from side to side

Muscles which when contracted push the tongue out

Hyoid bone

Sheath of skin

Muscles which pull the tongue in

The woodpecker's skeleton shows several special adaptations. The skull is particularly sturdy to withstand the force of the blows as it hammers with its beak; the leg bones are large and strong; the end of the spine curves downwards to enable the tail feathers to act as a prop. One toe of the woodpecker's foot is reversed to give a stronger grip.

Woodpecker

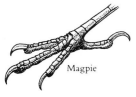

Magpie

The feathers in the woodpecker's tail are exceptionally stiff, and the outermost feathers are reduced in size, allowing the tail to act as a prop.

Chapter 1
Variation under Domestication

Causes of Variability

[*Darwin's idea that domestication can, in itself, cause greater variability to arise between individuals is now known to be wrong. The greater variability which is seen in domesticated plants and animals is the result of their not having been subjected to natural selection which, if the environment is stable and the species well adapted, tends to eliminate those which depart from the norm. Gene recombination and mutation are what give rise to variation (see p. 26), and these will occur at the same rate in the wild as under domestication. But in the wild, variations will usually be far more ruthlessly weeded out.*]

When we compare the individuals of the same variety of our older cultivated plants and animals, one of the first points which strikes us is that they generally differ more from each other than do the individuals of any one species or variety in a state of nature. And if we reflect on the vast diversity of the plants and animals which have been cultivated during all ages under the most different climates, we are driven to conclude that this great variability is due to our domestic productions having been raised under conditions of life not so uniform as, and somewhat different from, those to which the parent-species had been exposed under nature. It seems clear that organic beings must be exposed during several generations to new conditions to cause any great amount of variation; and that, when the organisation has once begun to vary, it continues varying for many generations. Our oldest cultivated plants, such as wheat, still yield new varieties: our oldest domesticated animals are still capable of modification.

The conditions of life appear to act in two ways –

directly on the organisation and indirectly by affecting the reproductive system. With respect to the direct action, there are two factors: namely, the nature of the organism, and the nature of the conditions. The former seems to be much the more important; for nearly similar variations sometimes arise under dissimilar conditions.

Effects of Habit and of the Use or Disuse of Parts

[*Characters acquired by an individual during its lifetime cannot be passed on to its offspring. Although Darwin wrongly thought that habit and the effects of use or disuse could be inherited, he did not regard the inheritance of such acquired characters as essential to his theory. In the first edition of* The Origin of Species *there was less emphasis on this subject, but Darwin later gave it more prominence to answer criticisms that there had not been enough time for so much evolution to have occurred merely by the accumulation of random variations.*]

Changed habits produce an inherited effect, and with animals the increased use or disuse of parts has a marked influence; thus in the domestic duck the bones of the wing weigh less and the bones of the leg more, in proportion to the whole skeleton, than do the same bones in the wild-duck; and this may be safely attributed to the domestic duck flying much less, and walking more, than its wild parents. The great and inherited development of the udders in cows and goats in countries where

'*Our oldest cultivated plants, such as wheat, still yield new varieties . . .*' Even Darwin might have been surprised to see the enormous changes effected by modern plant breeding. Some modern wheat is less than half as tall and yields three times as much grain as nineteenth-century varieties.

they are habitually milked, in comparison with these organs in other countries, is probably another instance of the effects of use. Not one of our domestic animals can be named which has not in some country drooping ears, and the view has been suggested that the drooping is due to the disuse of the muscles of the ear, from the animals being seldom much alarmed.

Correlated Variation; Inheritance

[*Since Darwin's time it has been shown that genes controlling different characters may be 'linked', that is, they are carried close together on the same chromosome and are therefore unlikely to be separated by crossing-over (see diagram on p. 24), so that they will tend to be inherited together. Moreover a single gene change can affect more than one character: such multiple effects of a single gene change are called 'pleiotropic effects'. There are often simple physiological links between characters: for example, a high level of the same hormone may affect several separate organs during development. Lastly, if we compare animals of similar form but of different sizes we find that some parts of the body change at different rates to others; these are called 'allometric changes'. These phenomena constitute most of what Darwin called the 'mysterious laws of correlation'. His other observations on inheritance can largely be explained in the light of modern knowledge about genetics.*]

Many laws regulate variation, some few of which can be dimly seen. I will here only allude to what may be called correlated variation. Breeders believe that long limbs are almost always accompanied by an elongated head. Some instances of correlation are quite whimsical: thus cats which are entirely white and have blue eyes are generally deaf; but this is confined to the males. Colour and constitutional peculiarities go together; it appears that white sheep and pigs are injured by certain plants, whilst dark-coloured individuals escape. Professor Wyman asked some farmers in Virginia how it was that all their pigs were black; they informed him that the pigs ate the paint-root, which

coloured their bones pink, and which caused the hoofs of all but the black varieties to drop off; and one of the 'crackers' (i.e. Virginia squatters) added, 'we select the black members of a litter for raising, as they alone have a good chance of living'. Hairless dogs have imperfect teeth; pigeons with feathered feet have skin between their outer toes; pigeons with short beaks have small feet, and those with long beaks large feet. Hence if man goes on selecting any peculiarity, he will almost certainly modify unintentionally other parts of the structure, owing to the mysterious laws of correlation.

The number and diversity of inheritable deviations, both those of slight and those of considerable physiological importance, are endless. No breeder doubts how strong is the tendency to inheritance; that like produces like is his fundamental belief: when amongst individuals, apparently exposed to the same conditions, any very rare deviation appears in the parent – say, once amongst several million individuals – and it reappears in the child, the mere doctrine of chances almost compels us to attribute its re-appearance to inheritance. Everyone must have heard of cases of albinism, prickly skin, hairy bodies, etc., appearing in several members of the same family. Thus, the inheritance of every character whatever is the rule, and non-inheritance the anomaly.

The laws governing inheritance are for the most part unknown. No one can say why the same peculiarity in different individuals of the same species is sometimes inherited and sometimes not; why the child often reverts in certain characters to its grandfather or grandmother or more remote ancestor; why a peculiarity is often transmitted from one sex to both sexes, or to one sex alone, more commonly but not exclusively to the like sex.

Character of Domestic Varieties; origin of Domestic Varieties from one or more Species

When we look to the hereditary varieties or races of our domestic animals and plants, and compare them with closely allied species, we generally perceive in each domestic race less uniformity of character than in true species. Domestic races often have a somewhat monstrous character; by which I mean, that, although differing from each other in several trifling respects, they often differ in an extreme degree in some one part, both when compared one with another, and more especially when compared with the species under nature to which they are nearest allied. With these exceptions, domestic races of the same species differ from each other in the same manner as do the closely allied species of the same genus in a state of nature, but the differences in most cases are less in degree.

In attempting to estimate the amount of structural difference between allied domestic races, we are soon involved in doubt, from not knowing whether they are descended from one or several parent species. In the case of most of our domesticated animals and plants, it is not possible to come to any definite conclusion. The argument mainly relied on by those who believe in multiple origins is that we find in the most ancient times, on the monuments of Egypt, and in the lake-habitations of Switzerland, much diversity in the breeds; and that some of these ancient breeds closely resemble those still existing. But this only shows that animals were domesticated at a much earlier period than has hitherto been supposed.

The lake-inhabitants of Switzerland cultivated several kinds of wheat and barley, the pea, the poppy for oil, and flax; and they possessed several domesticated animals. All this clearly shows that they had progressed considerably in civilisation; and this again implies a long-continued previous period of less advanced civilisation, during which the domesticated animals might have varied and given rise to distinct races. Since the discovery of flint tools in many parts of the world, all geologists believe that barbarian man existed at an enormously remote period; and at the present day there is hardly a tribe so barbarous as not to have domesticated at least the dog.

The origin of most of our domestic animals will probably for ever remain vague. But I have come to the conclusion that several wild species of Canidae have been tamed, and that their blood, in some cases mingled together, flows in the veins of our domestic breeds. In regard to sheep and goats I can form no decided opinion. From facts on the habits, voice, constitution, and structure of the humped Indian cattle, it is almost certain that they are descended from a different aboriginal stock from our European cattle; and some competent judges believe that these latter have had two or three wild progenitors. With respect to

A fresco from the tomb of Nebamun, Thebes, *c* 1400 B.C. In studying domestic varieties Darwin was hampered by not knowing whether only one species or more than one had originally been domesticated. Some biologists of his time believed that every distinctive variety was derived from a different wild species. They cited as evidence the diversity of breeds depicted by the Egyptians, and shown even by remains from Stone Age pile-dwelling settlements of the Swiss lakes. Darwin correctly argues that the existence of several different breeds in such early times merely indicates that domestication occurred at an even earlier period.

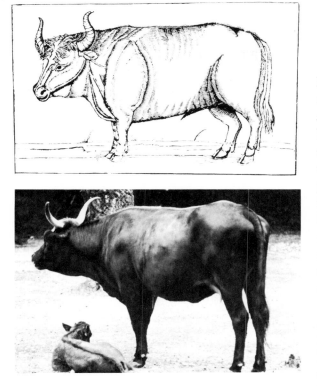

The origins of domesticated animals and plants are still the subject of extensive research, and much more is known than in Darwin's time. To take one example, cattle are now thought to be derived from a single species *Bos primigenius*, although various races of this species may have been independently domesticated, or crosses may have been made with other species after domestication to yield distinctive breeds. *Bos primigenius* was still alive in the wild until the seventeenth century; these wild cattle were known as aurochs. Using descriptions and drawings, such as the one made by Heberstain in 1549 *(above)*, attempts have been made to reconstitute the aurochs by crossing different breeds of cattle, and selecting auroch-like individuals for further breeding. A breed closely resembling the aurochs has been created, *(below)* and interestingly these animals show the ferocity and shyness characteristic of wild animals.

horses, I am doubtfully inclined to believe that all the races belong to the same species. Having kept nearly all the English breeds of the fowl alive, having bred and crossed them, and examined their skeletons, it appears to me almost certain that all are the descendants of the wild Indian fowl, *Gallus bankiva*.

The doctrine of the origin of our several domestic races from several aboriginal stocks has been carried to an absurd extreme by some authors. They believe that every race which breeds true has had its wild prototype. At this rate there must

have existed at least a score of species of wild cattle, as many sheep, and several goats, in Europe alone, and several even within Great Britain. One author believes that there formerly existed eleven wild species of sheep peculiar to Great Britain! When we bear in mind that Britain has now not one peculiar mammal, and France but few distinct from those of Germany, etc., but that each of these kingdoms possesses several peculiar breeds of cattle, sheep, etc., we must admit that many domestic breeds must have originated in Europe; for whence otherwise could they have been derived?

Breeds of the Domestic Pigeon, their Differences and Origin

Believing that it is always best to study some special group, I have taken up domestic pigeons. I have kept every breed which I could purchase or obtain, and have been kindly favoured with skins from several quarters of the world. I have associated with several eminent fanciers, and have been permitted to join two of the London Pigeon Clubs. The diversity of the breeds is something astonishing. Compare the English carrier and the short-faced tumbler, and see the wonderful difference in their beaks, entailing corresponding differences in their skulls. The carrier, especially the male bird, is also remarkable from the wonderful development of the carunculated skin about the head, and this is accompanied by greatly elongated eyelids, very large external orifices to the nostrils, and a wide gape of mouth. The short-faced tumbler has a beak in outline almost like that of a finch; and the common tumbler has the singular habit of flying at a great height in a compact flock, and tumbling in the air head over heels. The runt is a bird of great size; some of the sub-breeds have very long necks, others very long wings and tails, others singularly short tails. The barb is allied to the carrier, but, instead of a long beak, has a very short and broad one. The pouter has a much-elongated body, wings, and legs; and

Mr. Esquilant's Short-faced Baldheads.   Mr. W. Smith's White Pouters.   Mr. Wicking's Jacobin, Magpie, and Swallow.
Mr. Hayne's Carrier Cock.   Mr. Wicking's Magpie and Jacobin.
Mr. Harrison Weir's White Fantails.   Mr. Wicking's Brunswick and Nun.   Mr. Percival's Turbit.

PRIZE PIGEONS AT THE SHOW OF THE PHILO-PERISTERON SOCIETY, RECENTLY HELD IN FREEMASONS' HALL.

Fancy pigeon breeds as exhibited in 1864 by one of the London pigeon clubs to which Darwin belonged. All domesticated pigeons are descended from the rock dove, which closely resembles the feral pigeon familiar to town-dwellers. The extraordinary diversity of fancy pigeon breeds has been achieved through intensive selection by man.

its enormously developed crop, which it glories in inflating, may well excite astonishment and even laughter. The turbit has a short and conical beak, with a line of reversed feathers down the breast; and it has the habit of continually expanding the upper part of the oesophagus. The Jacobin has the feathers so much reversed along the back of the neck that they form a hood; and it has, proportionally to its size, elongated wing- and tail-feathers. The fantail has thirty or even forty tail-feathers, instead of the normal twelve or fourteen: these feathers are kept expanded, and are carried so erect that the head and tail touch.

In the skeletons of the several breeds, the development of the bones of the face differs enormously. The vertebrae vary in number; as does the number of the ribs, together with their relative breadth and the presence of processes. The number of the primary wing and caudal feathers, the relative length of the wing and tail, the relative length of the leg and foot, the development of skin between the toes, are all points of structure which are variable. The period at which the perfect plumage is acquired varies, as do the shape and size of the eggs, the manner of flight, and in some breeds the voice and disposition. Lastly, in certain breeds, the males and females have come to differ in a slight degree from each other.

Altogether at least a score of pigeons might be chosen which, if shown to an ornithologist, and he were told that they were wild birds, would certainly be ranked by him as well-defined species.

Moreover, I do not believe that any ornithologist would in this case place the English carrier, the short-faced tumbler, the runt, the barb, pouter, and fantail in the same genus; in each of these breeds several truly-inherited sub-breeds, or species, as he would call them, could be shown him.

Great as are the differences between the breeds of the pigeon, I am fully convinced that all are descended from the rock-pigeon [rock dove] *Colomba livia*. As several of the reasons which have led me to this belief are applicable in other cases, I will here briefly give them. If the several breeds are not varieties, and have not proceeded from the rock-pigeon, they must have descended from at least seven aboriginal stocks, for it is impossible to make the present domestic breeds by the crossing of any lesser number: how, for instance, could a pouter be produced by crossing unless one of the parent-stocks possessed the characteristic enormous crop? The supposed aboriginal stocks must all have been rock-pigeons, that is, they did not breed or willingly perch on trees. But beside *C. livia*, only two or three other species of rock-pigeons are known; and these have not any of the characters of the domestic breeds. Hence the supposed aboriginal stocks must have become extinct in the wild state. But birds breeding on precipices and good fliers are unlikely to be exterminated; and the common rock-pigeon has not been exterminated even on the smaller British islets. Again, it is difficult to get wild animals to breed freely under domestication; yet, on the hypothesis of the multiple origin of our pigeons, it must be assumed that at least seven species were so thoroughly domesticated in ancient times, as to be quite prolific under confinement.

The above-specified breeds, though agreeing generally with the wild rock-pigeon in constitution, habits, etc., are certainly highly abnormal in other parts; we look in vain through the whole family of Columbidae for a beak like that of the English carrier, short-faced tumbler, or barb, or for reversed feathers like those of the Jacobin.

Hence it must be assumed not only that half-civilised man succeeded in thoroughly domesticating several species, but that he picked out extraordinarily abnormal species; and further, that these species have since all become extinct. So many strange contingencies are improbable in the highest degree.

Some facts in regard to the colouring of pigeons deserve consideration. The rock-pigeon is of a slaty-blue, with white loins. The tail has a terminal dark bar, with the outer feathers externally edged at the base with white. The wings have two black bars. Some semi-domestic breeds have, besides the two black bars, the wings chequered with black. These several marks do not occur together in any other species of the whole family. Now, in every domestic breed all the above marks sometimes concur perfectly developed. Moreover, when birds belonging to two or more distinct breeds are crossed, none of which are blue or have any of the above-specified marks, the mongrel offspring are very apt suddenly to acquire these characters. [*See note on p. 101.*] We can understand these facts, on the well-known principle of reversion to ancestral characters, if all the domestic breeds are descended from the rock-pigeon.

Lastly, the mongrels from between all the breeds of the pigeon are perfectly fertile. Now hardly any cases have been ascertained with certainty of hybrids from two distinct species of animals being perfectly fertile.

From these several reasons, we may conclude that all our domestic breeds are descended from the rock-pigeon with its geographical sub-species. In favour of this view, I may add, firstly, that the wild *C. livia* has been found capable of domestication in Europe and India; and that it agrees in habits and a great number of points of structure with all domestic breeds. Secondly, although a carrier or tumbler differs immensely in certain characters from the rock-pigeon, yet by comparing the several sub-breeds of these two races, we can make, between them and the rock-pigeon, an almost perfect series; so we can in some other cases, but not with all the breeds. Thirdly, those characters which are mainly distinctive of each breed are in each eminently variable, for instance the wattle and length of beak of the carrier, the shortness of that of the tumbler, and the number of tail-feathers in the fantail.

I have discussed the probable origin of domestic pigeons at some length because when I first kept pigeons, well knowing how truly they breed, I felt as much difficulty in believing that since they had been domesticated they had all proceeded from a common parent, as any naturalist could in coming to a similar conclusion in regard to the many species of finches, or other groups of birds, in nature. May not those naturalists who admit that many of our domestic races are descended from the same parents not learn a lesson of caution, when they deride the idea of species in a state of nature being lineal descendants of other species?

Principles of Selection anciently followed, and their Effects

Let us now consider the steps by which domestic races have been produced. One of the most remarkable features in our domesticated races is that we see adaptation, not to the animal's or plant's own good, but to man's use or fancy. Some variations useful to him have probably arisen suddenly; many botanists believe that the Fuller's teasel, with its hooks, which cannot be rivalled by any mechanical contrivance, is only a variety of the wild *Dipsacus*; and this amount of change may have suddenly arisen in a seedling. But when we compare the dray-horse and race-horse, the various breeds of sheep fitted either for cultivated land or mountain pasture, with the wool of one breed good for one purpose, and that of another breed for another purpose; when we compare the host of agricultural, culinary, and flower-garden races of plants, most useful to man at different seasons and for different purposes, or so beautiful in his eyes, we must look farther than to mere variability.

'*. . . many botanists believe that the Fuller's teasel, with its hooks, which cannot be rivalled by any mechanical contrivance is only a variety of the wild Dipsacus; and this amount of change may have suddenly arisen in a seedling.*' **The difference between the wild teasel** (*left*) **and the Fuller's teasel** (*right*) **is probably due to a single mutation. The Fuller's teasel is still used today for raising the nap on woollen cloth.**

We cannot suppose that all the breeds were suddenly produced as perfect and as useful as we now see them. The key is man's power of accumulative selection: nature gives successive variations; man adds them up in certain directions useful to him.

The great power of this principle of selection is not hypothetical. Several of our eminent breeders have, even within a lifetime, modified to a large extent their breeds of cattle and sheep. Breeders habitually speak of an animal's organisation as something plastic, which they can model almost as they please. In Saxony the importance of the principle of selection in regard to merino sheep is so fully recognised, that men follow it as a trade: the sheep are placed on a table and studied three times at intervals of months, and the sheep are each time marked and classed, so that the very best may ultimately be selected for breeding. If selection consisted merely in separating some very distinct variety, and breeding from it, the principle would be so obvious as hardly to be worth notice; but its importance consists in the great effect produced by the accumulation in one direction, during successive generations, of differences absolutely inappreciable by an uneducated eye.

No one supposes that our choicest productions have been produced by a single variation from the aboriginal stock. We have proofs that this has not been so in several cases in which exact records have been kept; the steadily increasing size of the common gooseberry may be quoted. We see an astonishing improvement in many florists' flowers, when the flowers of the present day are compared with drawings made only twenty or thirty years ago. When a race of plants is once pretty well established, the seed-raisers do not pick out the best plants, but merely pull up the 'rogues', as they call the plants that deviate from the

'In regard to plants, there is another means of observing the accumulated effects of selection – namely, by observing the diversity of pods, tubers, or whatever part is valued in the kitchen garden, in comparison with the flowers of the same varieties . . .' **This applies equally well to related species such as the aubergine or egg plant** *Solanum melongena* **(left) and the potato** *Solanum tuberosum* **(right).**

proper standard. With animals this kind of selection is, in fact, likewise followed, for hardly anyone is so careless as to breed from his worst animals.

In regard to plants, there is another means of observing the accumulated effects of selection – namely, by observing the diversity of pods, tubers, or whatever part is valued in the kitchen-garden, in comparison with the flowers of the same varieties: and the diversity of fruit of the same species in the orchard, in comparison with the leaves and flowers of the same set of varieties. See how different the leaves of the cabbage are, and how extremely alike the flowers; how unlike the flowers of the heartsease [pansies] are, and how alike the leaves. It is not that the varieties which differ largely in some one point do not differ at all in other points; this is perhaps never the case. The law of correlated variation, the importance of which should never be overlooked, will ensure some differences; but as a general rule, the continued selection of slight variations, either in the leaves, the flowers, or the fruit, will produce races differing from each other chiefly in these characters.

It may be objected that the principle of selection has been reduced to methodical practice for scarcely more than three-quarters of a century. But it is very far from true that the principle is a modern discovery. In rude and barbarous periods of English history choice animals were often imported, and laws were passed to prevent their exportation: the destruction of horses under a certain size was ordered. The principal of selection I find distinctly given in an ancient Chinese encyclopaedia. Explicit rules are laid down by some of the Roman classical writers. From passages in Genesis, it is clear that the colour of domestic animals was at that early period attended to. Savages now sometimes cross their dogs with wild canine animals to improve the breed, and they formerly did so, as is attested by passages in Pliny. Livingstone states that good domestic breeds are highly valued by the negroes in the interior of Africa who have not associated with

Europeans. Some of these facts do not show actual selection, but they show that the breeding of domestic animals was carefully attended to in ancient times, and is now attended to by the lowest savages.

Unconscious Selection

Eminent breeders try by methodical selection, with a distinct object in view, to make a superior strain, but, for our purpose, a form of Selection which may be called Unconscious, and which results from everyone trying to possess and breed from the best individual animals, is more important. Thus, a man who intends keeping pointers naturally tries to get as good dogs as he can, and afterwards breeds from his best dogs, but he has no wish or expectation of permanently altering the breed. Nevertheless we may infer that this process, continued during centuries, would improve any breed. Slow and insensible changes of this kind can never be recognised unless actual measurements of the breeds have been made long ago, which may serve for comparison. In some cases, however, unchanged individuals of the same breed exist in less civilised districts, where the breed has been less improved. The King Charles's spaniel has probably been unconsciously modified to a large extent since the time of that monarch. The English pointer has been greatly changed within the last century, and the change has been effected unconsciously and gradually, and yet so effectually, that, though the old Spanish pointer certainly came from Spain, Mr. Borrow has not seen any native dog in Spain like our pointer.

By a similar process English racehorses have come to surpass in fleetness and size the parent Arabs, so that the latter, by the regulations for the Goodwood Races, are favoured in the weights which they carry. By comparing the accounts given in various old treatises we can trace the stages through which carrier- and tumbler-pigeons have insensibly come to differ so greatly from the rock-pigeon.

If there exist savages so barbarous as never to think of the inherited character of their domestic animals, yet any one animal particularly useful to them would be carefully preserved during famines, and such choice animals would thus generally leave more offspring than the inferior ones; so in this case there would be a kind of unconscious selection going on. We see the value set on animals even by the barbarians of Tierra del Fuego, by their killing and devouring their old women in times of dearth, as of less value than their dogs.

A native of Tierra del Fuego with his dog: the savagery of these people, whom Darwin encountered during the voyage of the *Beagle*, made a deep and lasting impression on him. One of the sailors was told by a Fuegian that during a famine they would kill and eat their old women but not their dogs.

In plants the same gradual process of improvement, through the occasional preservation of the best individuals, may plainly be recognised in the increased size and beauty in the varieties of the heartsease, rose, dahlia and other plants, when compared with older varieties or their parent-stocks. No one would expect to raise a first-rate melting pear from the seed of the wild pear, though he might succeed from a poor seedling growing wild if it had come from a garden-stock. The pear, though cultivated in classical times, appears to have been a fruit of very inferior quality. I have seen great surprise expressed at the skill of gardeners, in having produced such splendid results from such poor materials; but the art has been simple and followed almost unconsciously. It has consisted in always cultivating the best known variety, sowing its seeds, and, when a slightly better variety chanced to appear, selecting it, and so onwards.

A large amount of change, thus slowly and unconsciously accumulated, explains why in a number of cases we cannot recognise the wild parent-stocks of the plants which have been longest cultivated in our gardens. If it has taken centuries to improve most of our plants up to their present standard of usefulness to man, we can understand how it is that neither Australia, the Cape of Good Hope, nor any other region inhabited by quite uncivilised man, has afforded us a single plant worth culture. It is not that these countries, so rich in species, do not possess the aboriginal stocks of any useful plants, but that the native plants have not been improved by continued selection up to a standard comparable with that acquired by the plants in countries anciently civilised.

On the view here given of the important part which selection by man has played, it becomes obvious how it is that our domestic races show adaptation in their structure or habits to man's wants or fancies. We can further understand the frequently abnormal character of our domestic

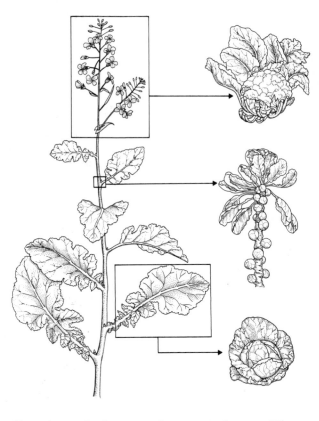

'Domestic races often have a somewhat monstrous character ...' The cabbage, brussels sprout, and cauliflower are all varieties of the same species, *Brassica oleracea*. The wild form of this species is shown on the left. Different parts of the plant have been developed by man's selection. In the cabbage it is the leaves, in the brussels sprout the side buds, and in the cauliflower the flowerhead, which form the edible part and are over-developed in size. In spite of the enormous visible differences, if allowed to flower, all three can still be cross-pollinated.

races, and likewise their differences being so great in external characters, and relatively so slight in internal parts. Man can hardly select any deviation of structure excepting such as is externally visible; and indeed he rarely cares for what is internal. He can never act by selection, excepting on variations which are first given to him in some slight degree by nature. No man would ever try to make a fantail till he saw a pigeon with a tail developed in some slight degree in an unusual manner; and the more unusual any character, the more likely it would be to catch his attention. But to use such an expression as trying to make a fantail, is, in most cases, utterly incorrect. The man who first selected a pigeon with a slightly larger tail never dreamed what the descendants of that pigeon would become through long-continued, partly unconscious and partly methodical selection.

Nor would some great deviation of structure be necessary to catch the fancier's eye: he perceives extremely small differences. Nor must the value which would formerly have been set on any slight differences in the individuals of the same species, be judged of by the value which is now set on them, after several breeds have fairly been established.

Circumstances favourable to Man's Power of Selection

A high degree of variability is obviously favourable to man's power of selection as freely giving the materials for selection to work on. As variations manifestly useful or pleasing to man appear only occasionally, the chance of their appearance will be much increased by a large number of individuals being kept. Hence, number is of the highest importance for success. Nurserymen, from keeping large stocks of the same plant, are generally far more successful than amateurs in raising new varieties. A large number of individuals can be reared only where the conditions for propagation are favourable. When the individuals are scanty all will be allowed to breed, whatever their quality, and this will effectively prevent selection. But most importantly the animal or plant should be so highly valued by man, that attention is paid to even the slightest deviations in its qualities or structure. Unless such attention be paid, nothing can be effected. The strawberry has always varied since it was cultivated, but the slight varieties were formerly neglected. As soon, however, as gardeners picked out individual plants with slightly larger, earlier, or better fruit, and raised seedlings from them, and again picked out the best seedlings and bred from them, then (with some aid by crossing distinct species) those many admirable varieties of the strawberry were

raised which have appeared during the last half-century.

With animals, preventing crosses facilitates the formation of new races. In this respect enclosure of the land plays a part. Pigeons can be mated for life; thus many races may be improved and kept true, though mingled in the same aviary; and this circumstance must have favoured the formation of new breeds. Pigeons can be propagated in great numbers and at a very quick rate, and inferior birds may be freely rejected, as when killed they serve for food. On the other hand cats, from their nocturnal rambling habits, cannot be easily matched, and, although so much valued by women and children, we rarely see a distinct breed long kept up. The rarity of distinct breeds of cat, donkey, etc., may be attributed in main part to selection not having been brought into play; in cats, from the difficulty in pairing them; in donkeys, from only a few being kept by poor people, and little attention paid to their breeding; for recently in certain parts of Spain and the United States this animal has been surprisingly improved by careful selection.

To sum up on the origin of our domestic races of animals and plants. Changed conditions of life are of the highest importance in causing variability, which is governed by many unknown laws, of which correlated variation is probably the most important. Something, but how much we do not know, may be attributed to the definite action of the conditions of life. Some, perhaps a great, effect may be attributed to the increased use or disuse of parts. In some cases the intercrossing of aboriginally distinct species appears to have played an important part in the origin of our breeds, but the importance of crossing has been much exaggerated. Over all these causes of Change, the accumulative action of Selection, whether applied methodically and quickly, or unconsciously and slowly but more efficiently, seems to have been the predominant Power.

Chapter 2
Variation under Nature

Before applying the principles arrived at in the last chapter to organic beings in a state of nature, we must briefly discuss whether these latter are subject to any variation. Here I shall not discuss the various definitions which have been given of the term species. Generally the term includes the unknown element of a distinct act of creation. The term 'variety' is almost equally difficult to define; but here community of descent is almost universally implied, though it can rarely be proved. We have also what are called monstrosities; but they graduate into varieties. By a monstrosity is meant some sudden and considerable deviation of structure, generally injurious or not useful to the species, such as we occasionally see in our domestic productions, especially plants. It may be doubted whether these are ever permanently propagated in a state of nature.

Individual Differences

The many slight differences which appear in the offspring from the same parents may be called individual differences. No one supposes that all the individuals of the same species are cast in the same actual mould. These individual differences are of the highest importance for us, for they are often inherited; they thus afford materials for natural selection to act on and accumulate, in the same manner as man accumulates individual differences in his domesticated productions. These individual differences generally affect unimportant parts, but parts which must be called important sometimes vary in individuals of the same species.

Individuals of the same species often present great differences of structure, independently of variation, as in the two sexes of various animals, in the two or three castes of sterile females or workers amongst insects, and in the immature and larval states of many of the lower animals. There are, also, cases of dimorphism and trimorphism, both with animals and plants. Thus, Mr. Wallace has shown that the females of certain species of butterflies, in the Malayan Archipelago, regularly appear under two or even three conspicuously distinct forms not connected by intermediate varieties. Although

The Malaysian butterfly *Papilio memnon agenor* observed by Alfred Wallace to show several distinct forms in the female. The male is shown on the left, three females on the right.

in most cases the two or three forms, both with animals and plants, are not now connected by intermediate gradations, it is probable that they were once thus connected. Mr. Wallace describes a certain butterfly which presents in the same island a great range of varieties connected by intermediate links, and the extreme links of the chain closely resemble the two forms of an allied dimorphic species inhabiting another part of the Malay Archipelago. It certainly at first appears highly remarkable that the same female butterfly should produce at the same time three distinct female forms and a male. Nevertheless these cases are only exaggerations of the common fact that the female produces offspring of two sexes which sometimes differ from each other in a wonderful manner.

Doubtful Species

[*In this section Darwin first puts forward his argument, which continues throughout the book, that species develop from varieties. His opponents believed that each species had been separately created by God, but that varieties had arisen within these species by natural variation. Hence Darwin's overriding objective is to minimize the distinction between species and varieties. Modern biologists would define a species as a group of individuals all of which can potentially interbreed one with another. This is a working definition which holds good in the majority of cases. There are anomalies however, some of which Darwin deals with in Chapter 9, where he is using them in favour of his argument.*] The forms which possess in some considerable degree the character of species, but which are so closely similar to other forms, or are so closely linked to them by intermediate gradations, that naturalists do not like to rank them as distinct species, are in several respects the most important for us. Many of these closely allied forms have retained their characters for a long time; for as long, as far as we know, as have good and true species. Practically, when a naturalist can unite by means of intermediate links any two forms, he treats the one as a variety of the other; ranking the most common, as a variety of

A barren-ground caribou from Canada. The reindeer of northern Europe are slightly smaller but otherwise are very similar. The two are now generally considered to be varieties of the same species, but they were once classified as separate species. Taxonomy is continually changing in this way, and taxonomists, as Darwin points out, frequently disagree. For this reason Darwin argues that the distinction between species and varieties is '*entirely vague and arbitrary*'.

but sometimes the one first described, as the species, and the other as the variety. In very many cases, however, one form is ranked as a variety of another, not because the intermediate links have actually been found, but because analogy leads the observer to suppose either that they do now somewhere exist, or may formerly have existed. But few well-marked varieties can be named which have not been ranked as species by at least some competent judges.

Many years ago, when comparing the birds from the closely neighbouring islands of the Galapagos Archipelago, one with another, and with those from the American mainland, I was much struck how entirely vague and arbitrary is the distinction between species and varieties. On the islets of the little Madeira group there are many insects which are characterised as varieties in Mr. Wollaston's admirable work, but which would certainly be ranked as distinct species by many entomologists. Several experienced ornithologists consider our British red grouse as only a strongly marked race of a Norwegian species, whereas the greater number rank it as an undoubted species peculiar to Great Britain.

Some few naturalists maintain that animals never present varieties; but then these same naturalists rank the slightest difference as of specific value; and when the same identical form is met with in two distant countries, or in two geological formations, they believe that two distinct species are hidden under the same dress. The term species thus comes to be a mere useless abstraction, implying and assuming a separate act of creation.

Certainly no clear line of demarcation has as yet been drawn between species and sub-species — that is, the forms which in the opinion of some naturalists come very near to, but do not quite arrive at, the rank of species: or, again, between sub-species and well-marked varieties, or between lesser varieties and individual differences. These differences blend into each other by an insensible series; and a series impresses the mind with the idea of an actual passage.

Hence I look at individual differences as of the highest importance for us, as being the first steps towards such slight varieties as are barely thought worth recording. And I look at varieties which are in any degree more distinct and permanent as steps towards more strongly marked and permanent varieties; and at the latter, as leading to sub-species, and then to species. The passage from one stage to another may be safely attributed to the cumulative action of natural selection, hereafter to be explained. A well-marked variety may therefore be called an incipient species; but whether this belief is justifiable must be judged by the weight of the various facts and considerations to be given throughout this work. Not all varieties attain the rank of species. They may become extinct, or they may endure as varieties for very long periods.

From these remarks it will be seen that I look at the term species as one arbitrarily given, for the sake of convenience, to a set of individuals closely resembling each other, and that it does not essentially differ from the term variety, which is given to less distinct and more fluctuating forms.

Wide-ranging, much-diffused, and common Species vary most; species of the Larger Genera in each Country vary more frequently than the Species of the Smaller Genera

Plants which have very wide ranges generally present varieties, as they are exposed to diverse physical conditions, and come into competition with different sets of organic beings. Furthermore, in any limited country, the species which abound most in individuals, and the species which are most widely diffused within their own country oftenest give rise to varieties sufficiently well-marked to have been recorded in botanical works. And this, perhaps, might have been anticipated; for, as varieties, in order to become in any degree permanent, necessarily have to struggle with the other inhabitants of the country, the species which are already dominant will be the most likely to yield offspring, which, though in some slight degree modified, still inherit those advantages that enabled their parents to become dominant over their compatriots, that is, the forms having similar habits of life.

If the plants inhabiting a country be divided into two equal masses, all those in the larger genera (i.e. those including many species) being placed on one side, and all those in the smaller genera on the other side, the former will be found to include a somewhat larger number of the dominant species. For the mere fact of many species of the same genus inhabiting any country shows that there is something in the conditions of that country favourable to the genus; and, consequently, we might have expected to have found in the larger genera, or those including many species, a larger proportional number of dominant species.

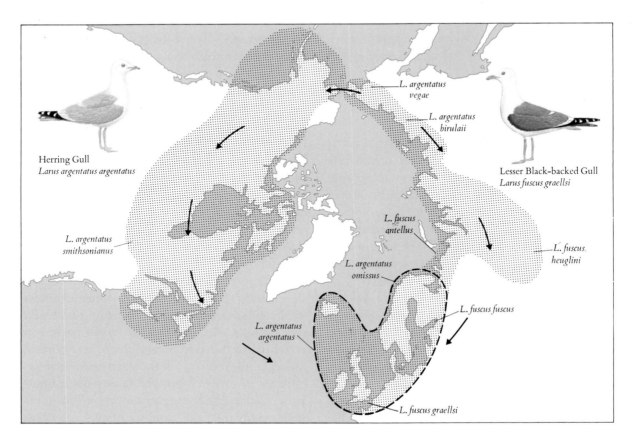

Herring Gull
Larus argentatus argentatus

L. argentatus smithsonianus

L. argentatus vegae

L. argentatus birulaii

Lesser Black-backed Gull
Larus fuscus graellsi

L. fuscus antellus

L. fuscus heuglini

L. argentatus omissus

L. argentatus argentatus

L. fuscus fuscus

L. fuscus graellsi

'Certainly no clear line of demarcation has yet been drawn between species and subspecies. These differences blend into each other by an insensible series; and a series impresses the mind with the idea of an actual passage.' Such a passage can be seen in the herring gull and lesser black-backed gull which form what is known as a ring species. Five subspecies of the former and four of the latter are recognised, forming a chain that goes around the North Pole. Each subspecies can breed with those on either side of it. The evidence strongly suggests that all are derived from an ancestral population that originated in Siberia. Members of this ancestral population have migrated in opposite directions, and at the same time evolved so that at various stages new subspecies can be identified. It would be reasonable to treat all the subspecies as belonging to the same species except that where the two ends of the chain overlap in northern Europe they do not interbreed except on very rare occasions. The gulls are therefore assigned to two species *Larus argentatus* and *Larus fuscus*, but the position of the dividing line between the species is clearly arbitrary: *L. argentatus birulaii* could quite well be assigned to *L. fuscus*, or *L. fuscus heuglini* to *L. argentatus*.

From looking at species as only well-defined varieties, I was also led to anticipate that the species of the larger genera would oftener present varieties than the species of the smaller genera; for wherever many closely related species (i.e. species of the same genus) have been formed, many varieties or incipient species ought, as a general rule, to be now forming. Where circumstances have been favourable for variation, we might expect that they would generally be still favourable to variation.

To test the truth of this anticipation I have arranged the plants of twelve countries, and the coleopterous insects of two districts, into two nearly equal masses, the species of the larger genera on one side, and those of the smaller genera on the other side; and a larger proportion of the species on the side of the larger genera presented varieties, than on the side of the smaller genera. Moreover, the species of the larger genera invariably present a larger average number of varieties than do the species of the small genera. These facts are of plain signification on the view that species are only strongly marked and permanent varieties; for wherever the manufactory of species has been active, we ought generally to find the manufactory still in action.

Many of the Species included within the Larger Genera resemble Varieties

Fries has remarked in regard to plants, and Westwood in regard to insects, that in large genera the amount of difference between the species is often exceedingly small. I have also consulted some sagacious and experienced observers, and they concur in this view. In this respect, therefore, the species of the larger genera resemble varieties more than do the species of the smaller genera.

Or the case may be put in another way, and it may be said that in the larger genera, in which a number of varieties greater than the average are now manufacturing, many of the species already manufactured still to a certain extent resemble varieties, for they differ from each other by less than the usual amount of difference. Moreover, the species of the larger genera are related to each other in the same manner as varieties are related to each other. Little groups of species are generally clustered like satellites around other species. And what are varieties but groups of forms, unequally related to each other, and clustered round their parent-species?

Summary

We see that it is the most flourishing species of the larger genera which on average yield the greatest number of varieties; and varieties, as we shall hereafter see, tend to become converted into new and distinct species. Thus the larger genera tend to become larger; and throughout nature the forms of life which are now dominant tend to become still more dominant by leaving many modified and dominant descendants. But by steps hereafter to be explained, the larger genera also tend to break up into smaller genera. And thus, the forms of life throughout the universe become divided into groups subordinate to groups.

Chapter 3
Struggle for Existence

Before entering on the subject of this chapter, I must show how the struggle for existence bears on Natural Selection. The mere existence of individual variability, though necessary as the foundation for the work, helps us but little in understanding how species arise in nature. How have all those exquisite adaptations of one part of the organisation to another part, and to the conditions of life, and of one organic being to another being, been perfected? We see these beautiful coadaptations most plainly in the woodpecker and the mistletoe; and only a little less plainly in the humblest parasite which clings to the hairs of a quadruped or feathers of a bird; in the structure of the beetle which dives through the water; in the plumed seed which is wafted by the gentlest breeze.

Owing to the struggle for life, variations, if they be in any degree profitable to the individuals of a species, in their infinitely complex relations to other organic beings and to their physical conditions of life, will tend to the preservation of such individuals, and will generally be inherited by the offspring. The offspring, also, will thus have a better chance of surviving, for, of the many individuals of any species which are born, but a small number can survive. I have called this principle, by which each slight variation, if useful, is preserved, by the term Natural Selection, in order to mark its relation to man's power of selection. But the expression often used by Mr. Herbert Spencer of the Survival of the Fittest is more accurate, and is sometimes equally convenient.

Man can by selection adapt organic beings to his own uses, through the accumulation of slight variations given to him by the hand of Nature. But Natural Selection, as we shall hereafter see, is a power incessantly ready for action, and is as immeasurably superior to man's feeble efforts, as the works of Nature are to those of Art.

Nothing is easier than to admit in words the truth of the universal struggle for life, or more difficult than constantly to bear this conclusion in mind. We behold the face of nature bright with gladness, we often see super-abundance of food; we do not see, or we forget, that the birds which are idly singing round us mostly live on insects or seeds, and are thus constantly destroying life; or we forget how largely these songsters, or their eggs, or their nestlings, are destroyed by birds and beasts of prey; we do not always bear in mind that, though food may be now super-abundant, it is not so at all seasons of each recurring year.

The Term, Struggle for Existence, used in a large sense

I use this term in a large and metaphorical sense including dependence of one being on another, and including (which is more important) not only the life of the individual, but success in leaving progeny. Two canine animals, in a time of dearth, may be truly said to struggle with each other which shall get food and live. But a plant on the edge of a desert is said to struggle for life against the drought, though more properly it should be said to be dependent on the moisture. A plant which annually produces a thousand seeds, of which only one on an average comes to maturity, may be more truly said to struggle with the plants of the same and other kinds which already clothe the ground. The mistletoe is dependent on the apple and a few other trees, but can only in a far-fetched sense be said to struggle with these trees, for if too many of these parasites grow on the same tree it dies. But several seedling mistletoes, growing close together on the same branch, may more truly be said to struggle with each other. As the mistletoe is disseminated by birds, its existence depends on them; and it may metaphorically be said to struggle with other fruit-bearing plants, in tempting the birds to devour and thus disseminate its seeds. In these several senses, I use the general term of Struggle for Existence.

Geometrical Ratio of Increase

A struggle for existence inevitably follows from the high rate at which all organic beings tend to increase. Every being, which during its natural lifetime produces several eggs or seeds, must suffer destruction during some period of its life, otherwise, on the principle of geometrical increase, its numbers would quickly become so inordinately great that no country could support the product. Hence, as more individuals are produced than can possibly survive, there must in every case be a struggle for existence, either one individual with another of the same species, or with the individuals of distinct species, or with the physical conditions of life. It is the doctrine of Malthus applied to the whole animal and vegetable kingdoms. Although some species may be now increasing in numbers, all cannot do so, for the world would not hold them.

There is no exception to the rule that every organic being naturally increases at so high a rate that, if not destroyed, the earth would soon be covered by the progeny of a single pair. Even slow-breeding man has doubled in twenty-five years, and at this rate, in less than a thousand years there would literally not be standing-room for his progeny. Linnaeus has calculated that if an annual plant produced only two seeds – and there is no plant so unproductive – and their seedlings next year produced two, and so on, then in twenty

years there would be a million plants.

But we have better evidence on this subject from our domestic animals of many kinds which have run wild in several parts of the world; if the statements of the rate of increase of slow-breeding cattle and horses in South America and Australia had not been well authenticated, they would have been incredible. So it is with plants; the cardoon and a tall thistle, which are now the commonest over the wide plains of La Plata, clothing square leagues of surface almost to the exclusion of every other plant, have been introduced from Europe; and there are plants which now range in India from Cape Comorin to the Himalayas, which have been imported from America. In such cases, no one supposes that the fertility of the animals or plants has been suddenly increased in any sensible degree. The obvious explanation is that the conditions of life have been highly favourable, and that there has

'... of the many individuals of any species which are born, but a small number can survive.' Of these young tadpoles only a small percentage will survive to an age when they can themselves produce offspring.

'A plant which annually produces a thousand seeds, of which only one on an average comes to maturity, may be said to struggle with the plants of the same and other kinds which already clothe the ground.' **Beech seedlings germinating on the floor of a beech wood. Of those shown it is unlikely that more than one will grow into a tree and itself produce seeds. There will be intense competition for light and moisture and strong seedlings with rapid growth and efficient root systems will have an advantage.**

rapidly to increase in number. But the real importance of a large number of eggs or seeds is to make up for much destruction at some period of life; and this period in the great majority of cases is an early one. If an animal can protect its own eggs or young, a small number may be produced, and yet the average stock be fully kept up; but if many eggs or young are destroyed, many must be produced, or the species will become extinct. So that, in all cases, the average number of any animal or plant depends only indirectly on the number of its eggs or seeds.

In looking at Nature, it is most necessary never to forget that every single organic being may be said to be striving to the utmost to increase in numbers; that each lives by a struggle at some period of its life; that heavy destruction inevitably falls either on the young or old, during each generation or at recurrent intervals. Lighten any check, mitigate the destruction ever so little, and the number of the

'Two canine animals in a time of dearth, may be truly said to struggle with each other which shall get food and live.' **Timber wolves try to take away a deer killed by a grizzly bear.**

consequently been less destruction of the old and young, and that nearly all the young have been enabled to breed. Their geometrical ratio of increase simply explains their extraordinarily rapid increase and wide diffusion in their new homes.

The only difference between organisms which annually produce eggs or seeds by the thousand, and those which produce extremely few, is that the slow-breeders would require a few more years to people, under favourable conditions, a whole district, let it be ever so large. The Fulmar lays but one egg, yet it is believed to be the most numerous bird in the world. One fly deposits hundreds of eggs, and another, like the *hippobosca*, a single one; but this difference does not determine how many individuals of the two species can be supported in a district. A large number of eggs is of some importance to those species which depend on a fluctuating amount of food, for it allows them

'... *a plant on the edge of a desert is said to struggle for life against the drought, though more properly it should be said to be dependent on the moisture.*' **Death Valley, California.**

species will almost instantaneously increase to any amount.

Nature of the Checks to Increase

The causes which check the natural increase of each species are most obscure. Eggs or very young animals seem generally to suffer most, but this is not invariably the case. With plants there is a vast destruction of seeds, but it appears that the seedlings suffer most from germinating in ground already thickly stocked. Seedlings, also, are destroyed in vast numbers by various enemies; for instance, on a piece of ground three feet long and two wide, dug and cleared, and where there could be no choking from other plants, I marked all the seedlings of our native weeds as they came up, and out of 357 no less than 295 were destroyed, chiefly by slugs and insects. If turf which has long been mown or browsed be let to grow, the more vigor-

ous plants gradually kill the less vigorous, though fully grown plants.

The amount of food for each species gives the extreme limit to which each can increase; but very frequently it is the serving as prey to other animals which determines the average numbers of a species. Thus there seems to be little doubt that the stock of partridges, grouse, and hares on any large estate depends chiefly on the destruction of vermin. If not one head of game were shot during the next twenty years in England, and if no vermin were destroyed, there would be less game than at present, although hundreds of thousands of game animals are now annually shot. On the other hand, in some cases, as with the elephant, none are destroyed by beasts of prey; for even the tiger in India most rarely dares to attack a young elephant protected by its dam.

Climate plays an important part in determining the average numbers of a species, and periodical seasons of extreme cold or drought seem to be the most effective of all checks. I estimated that the winter of 1854–5 destroyed four-fifths of the birds in my own grounds. In so far as climate chiefly acts in reducing food, it brings on the most severe struggle between the individuals, whether of the same or of distinct species, which subsist on the same kind of food. Even when climate, for instance extreme cold, acts directly, it will be the least vigorous individuals, or those which have got least food through the advancing winter, which will suffer most. When we travel southward, or from a damp region to a dry, we invariably see some species gradually disappearing; and the change of climate being conspicuous, we are tempted to attribute the whole effect to its direct action. But we may feel sure that the cause lies quite as much in other species being favoured, as in this one being hurt. So it is when we travel northward, but in a somewhat lesser degree, for the number of species of all kinds, and therefore of competitors, decreases northwards; hence in going northwards, or in ascending a mountain, we far oftener meet with stunted forms, due to the *directly* injurious action of climate, than we do in proceeding southwards or in descending a mountain. When we reach the Arctic regions, or snow-capped summits, or absolute deserts, the struggle for life is almost exclusively with the elements.

That climate acts in main part indirectly by favouring other species, we clearly see in the prodigious number of plants which in our gardens can perfectly well endure our climate, but which never become naturalised, for they cannot compete with our native plants nor resist destruction by our native animals.

When a species increases inordinately in numbers in a small tract, epidemics often ensue; and here we have a limiting check independent of the struggle for life. But even some of these so-called epidemics are due to parasitic worms, which have through facility of diffusion amongst the crowded animals been disproportionally favoured: and here comes in a sort of struggle between the parasite and its prey.

On the other hand, in many cases, a large stock of individuals of the same species is necessary for its preservation. Thus we can easily raise plenty of corn and rape-seed in our fields, because the seeds are in great excess compared with the number of birds which feed on them; nor can the birds, though having a superabundance of food at this one season, increase in number proportionally to the supply of seed, as their numbers are checked during winter; but anyone who has tried knows how troublesome it is to get seed from a few wheat or other such plants in a garden. This explains why very rare plants are sometimes extremely abundant in the few spots where they do exist. For in such cases, a plant could exist only where the conditions of its life were so favourable that many could exist together, and thus save the species from utter destruction.

Complex Relations of all Animals and Plants to each other in the Struggle for Existence

Many cases show how complex and unexpected are the checks and relations between organic beings, which have to struggle together in the same country. I investigated a large and extremely barren heath, which had never been touched by the hand of man; several hundred acres of exactly the same nature had been enclosed twenty-five years previously and planted with Scotch fir. The change in the native vegetation of the planted part was most remarkable, more than is generally seen in passing from one quite different soil to another; not only the proportional numbers of the heath-plants were wholly changed, but twelve species of plants (not counting grasses and sedges) flourished in the plantations which could not be found on the heath. The effect on the insects must have been still greater, for six insectivorous birds were very common in the plantations which were not to be seen on the heath; and the heath was frequented by two or three distinct insectivorous birds. Here we see how potent has been the introduction of a single tree.

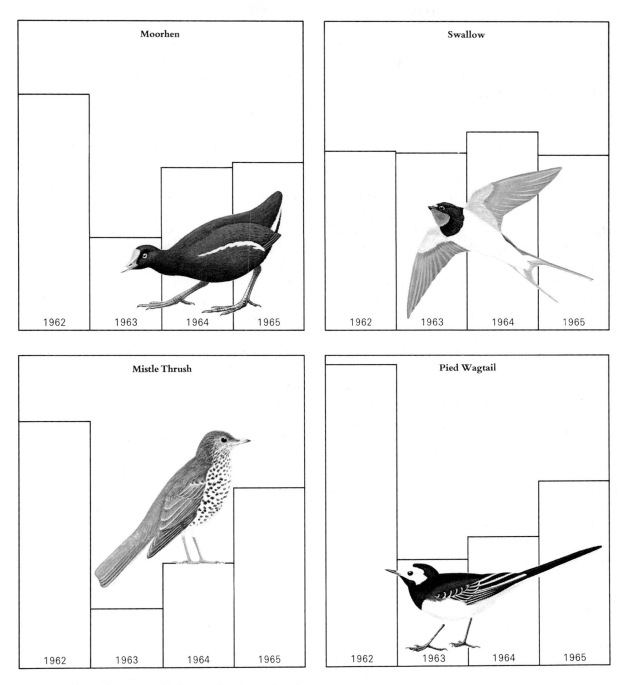

The effect of a harsh winter on bird populations in Britain. The
diagrams show the population levels during the summer of
each year from 1962 to 1965. The effect of the severe winter of
1962–3 in reducing the numbers of the moorhen, mistle thrush
and pied wagtail is evident. Migratory birds such as the
swallow were unaffected.

'When a species increases inordinately in numbers in a small tract, epidemics often ensue; and here we have a limiting check independent of the struggle for life.' This applies as much to man as to other species. All urban populations were threatened by outbreaks of epidemic diseases up until the twentieth century when improved public health measures led to their control. In Darwin's time the overcrowding in London was particularly acute, as shown by this view over London of 1860, and there were frequent typhus epidemics.

How important an element enclosure is, I plainly saw near Farnham, in Surrey. Here there are extensive heaths, with a few clumps of old Scotch firs on the hill-tops: within the last ten years large spaces have been enclosed, and self-sown firs are now springing up in multitudes, so close together that all cannot live. When I ascertained that these young trees had not been planted, I was so surprised at their numbers that I examined hundreds of acres of the unenclosed heath, and literally I could not see a single Scotch fir, except the old planted clumps. But on looking closely between the stems of the heath, I found a multitude of seedlings which had been perpetually browsed down by the cattle. In one square yard I counted thirty-two little trees; and one of them, with twenty-six rings of growth, had, during many years, tried to raise its head above the stems of the heath, and had failed. No wonder that, as soon as the land was enclosed, it became thickly clothed with vigorously growing young firs.

Here we see that cattle absolutely determine the existence of the Scotch fir; but in Paraguay insects determine the existence of cattle. For here neither cattle nor horses nor dogs have ever run wild, though they swarm southward and northward in a feral state. This is caused by the greater number in Paraguay of a certain fly, which lays its eggs in the navels of these animals when first born. The increase of these flies must be habitually checked by some means, probably by other parasitic insects. Hence, if certain insectivorous birds were to decrease in Paraguay, the parasitic insects would probably increase; and this would lessen the number of the navel-frequenting flies – then cattle and horses would become feral, and this would greatly alter the vegetation: this again would largely affect the insects; and this the insectivorous birds, and so onwards in ever-increasing circles of complexity. Not that under nature the relations will ever be as simple as this. Battle within battle must be continually recurring; and yet in the long-run the forces are so nicely balanced that the face of nature remains for long periods of time uniform, though assuredly the merest trifle would give the victory to one being over another. Nevertheless, so profound is our ignorance and so high our presumption that we marvel when we hear of the extinction of an organic being; and as we do not see the cause, we invoke cataclysms to desolate the world, or invent laws on the duration of the forms of life!

I am tempted to give one more instance showing how plants and animals are bound together by a web of complex relations. I find from experiments that humble-bees [bumble-bees] are almost indispensable to the fertilisation of the heartsease [wild pansy] and some kinds of clover. Humble-bees alone visit red clover, as other bees cannot reach the nectar. Hence we may infer that, if the whole genus of humble-bees became extinct or very rare in England, the heartsease and red clover would become very rare, or wholly disappear. The number of humble-bees in any district depends in a great measure on the number of field-mice, which destroy their combs and nests. Now the number of mice is largely dependent, as everyone knows, on the number of cats; and Col. Newman says, 'Near villages and small towns I have found

the nest of humble-bees more numerous than elsewhere, which I attribute to the number of cats that destroy the mice.' Hence it is quite credible that the presence of feline animals in large numbers might determine, through the intervention first of mice and then of bees, the frequency of certain flowers in that district!

In the case of every species, many different checks probably come into play, concurring in determining the average number or even the existence of the species. When we look at the plants clothing an entangled bank, we are tempted to attribute their proportional numbers and kinds to chance. But how false a view is this! Everyone has heard that when an American forest is cut down, a very different vegetation springs up; but it has been observed that ancient Indian ruins in the Southern United States, formerly cleared of trees, now display the same diversity and kinds as in the surrounding virgin forest. What a struggle must have gone on during long centuries between the several kinds of trees, each annually scattering its seeds by the thousand; what war between insect and insect – between insects, snails and other animals with birds and beasts of prey – all striving to increase, all feeding on each other, or on the trees, their seeds and seedlings, or on the other plants which first clothed the ground and thus checked the growth of the trees! Throw up a handful of feathers, and all fall to the ground according to definite laws; but how simple is the problem where each shall fall compared to that of the action and reaction of the innumerable plants and animals which have determined, in the course of centuries, the proportional numbers and kinds of trees now growing on the old Indian ruins!

The dependency of one organic being on another, as of a parasite on its prey, lies generally between beings remote in the scale of nature. But the struggle for existence will be most severe between individuals of the same species, for they require the same food and are exposed to the same dangers. In the case of varieties of the same species, the struggle will generally be almost equally severe: for instance, if several varieties of wheat be sown together, and the mixed seed be resown, some of the varieties which best suit the soil or climate, or are naturally the most fertile, will beat the others and so yield more seed, and will consequently in a few years supplant the other varieties. So again with sheep: certain mountain varieties will starve out others, so that they cannot be kept together. The same result has followed from keeping together different varieties of the medicinal leech.

Struggle for Life most severe between Individuals and Varieties of the same Species

As the species of the same genus usually have, though by no means invariably, much similarity in habits and constitution, and always in structure, the struggle will generally be more severe between them, if they come into competition with each other, than between the species of distinct genera. How frequently we hear of one species of rat taking the place of another species under the most different climates! In Australia the imported hive-bee is rapidly exterminating the small, stingless, native bee. We can dimly see why the competition should be most severe between allied forms, which fill nearly the same place in the economy of nature; but probably in no one case could we precisely say why one species has been so victorious over another in the great battle of life.

The structure of every organic being is related, in the most essential manner, to that of all other organic beings with which it comes into competition, from which it has to escape, or on which it preys. This is obvious in the teeth and talons of the tiger; and in the legs and claws of the parasite which clings to the hair on the tiger's body. But in the beautifully plumed seed of the dandelion, and in the flattened and fringed legs of the water-beetle, the relation seems at first confined to the elements of air and water. Yet the advantage of plumed seeds no doubt stands in the closest relation to the land being already thickly clothed with other plants; so that the seeds may be widely distributed and fall on unoccupied ground. In the water-beetle, the structure of its legs, so well adapted for diving, allows it to compete with other aquatic insects, to hunt for its own prey, and to escape serving as prey to other animals.

The store of nutriment laid up within the seeds of many plants seems at first sight to have no sort of relation to other plants. But from the strong growth of young plants produced from such seeds as peas and beans when sown in the midst of long grass, it may be suspected that the chief use of the nutriment in the seed is to favour the growth of the seedlings, whilst struggling with other plants growing vigorously all around.

We should keep steadily in mind that each organic being is striving to increase in a geometrical ratio; that each at some period of its life, during some season of the year, during each generation or at intervals, has to struggle for life and to suffer great destruction. When we reflect on this struggle, we may console ourselves with the full belief that the war of nature is not incessant, that no fear is felt, that death is generally prompt, and that the vigorous, the healthy, and the happy survive and multiply.

Chapter 4

Natural Selection;
or the Survival of the Fittest

How will the struggle for existence act in regard to variation? Can the principle of selection, so potent in the hands of man, apply under nature? Let it be borne in mind how infinitely complex and close-fitting are the mutual relations of all beings to each other and to their physical conditions of life; and consequently what diversities of structure might be of use under changing conditions. Can it, then, be thought improbable, seeing that variations useful to man have undoubtedly occurred, that other variations useful in some way to each being in the battle of life should occur in the course of many successive generations? If such do occur, can we doubt (remembering that many more individuals are born than can possibly survive) that individuals having any advantage, however slight, over others, would have the best chance of surviving and of procreating? On the other hand, we may feel sure that any variation in the least degree injurious would be rigidly destroyed. This preservation of favourable individual differences and variations, and the destruction of those which are injurious, I have called Natural Selection, or the Survival of the Fittest. Variations neither useful nor injurious would not be affected by natural selection, and would be left either a fluctuating element, as perhaps we see in certain polymorphic species, or would ultimately become fixed.

Several writers have misapprehended or objected to the term Natural Selection. Some have even imagined that natural selection induces variability, whereas it implies only the preservation of such variations as arise and are beneficial to the being under its conditions of life. It has been said

that I speak of natural selection as an active power or Deity; but who objects to an author speaking of the attraction of gravity as ruling the movements of the planets? So again it is difficult to avoid personifying the word Nature; but I mean by Nature only the aggregate action and product of many natural laws, and by laws the sequence of events as ascertained by us.

We shall best understand the probable course of natural selection by taking a country undergoing some slight physical change, for instance, of climate. The proportional numbers of its inhabitants will almost immediately undergo a change, and some species will probably become extinct. We may conclude, from what we have seen of the intimate and complex manner in which the inhabitants of each country are bound together, that any change in the numerical proportions of the inhabitants would seriously affect the others. If the country were open on its borders, new forms would immigrate, and this would likewise seriously disturb the relations of some of the former inhabitants. But in the case of an island, or of a country into which new forms could not freely enter, places in the economy of nature would assuredly be better filled up if some of the original inhabitants were in some manner modified. Slight modifications, which in any way favoured the individuals of any species, by better adapting them to their altered conditions, would tend to be preserved; and natural selection would have free scope for the work of improvement.

Unless profitable variations occur, natural selection can do nothing, but under the term of 'variations', it must never be forgotten that mere

Akiapolaau
Hawaii

Striped opossum
New Guinea

Woodpecker finch
Galapagos Islands

Aye–aye
Madagascar

Huia
New Zealand

'But in the case of an island, or of a country into which new forms could not freely enter, places in the economy of nature would assuredly be better filled up if some of the original inhabitants were in some manner modified.' On islands where woodpeckers are not found, other animals have evolved to fill their ecological niche. The woodpecker's specialities are its strong beak for excavating holes in wood, and its long, thin, flexible tongue for extracting insects from the crevices of the wood. Each of the animals pictured here has evolved special structures that simulate the woodpecker's beak and tongue. The aye-aye and the striped opossum have strong fingers with sharp claws for scratching the wood away, and one exceptionally long, thin finger for spearing insects. The woodpecker finch hammers the wood with its beak and uses a cactus spine held in its beak for probing. The akiapolaau, a species of honeycreeper, uses its long, curved upper beak for prising out insects; this upper beak has to be held open, out of the way, while the strong, straight lower beak is used to excavate the wood. The huia, an extinct bird, divided the labour between the sexes: the male had a sturdy beak for hammering, while the female had a long, curved beak for picking out the insects.

individual differences are included. Nor is any great change of climate or unusual degree of isolation to check immigration necessary in order that new and unoccupied places should be left for natural selection to fill. For as all the inhabitants of each country are struggling together with nicely balanced forces, extremely slight modifications in one species would often give it an advantage; still further modifications of the same kind would often further increase the advantage, as long as the species continued under the same conditions of life. No country can be named in which all the native inhabitants are so perfectly adapted that none of them could be still better adapted or improved; for in all countries the natives have allowed some foreigners to take firm possession of the land. Thus the natives might have been modified with advantage, so as to have better resisted the intruders.

As man can produce a great result by selection, what may not natural selection effect? Man can act only on external and visible characters: Nature, if I may be allowed to personify the natural preservation or survival of the fittest, cares nothing for appearances, except in so far as they are useful to any being. She can act on every internal organ, on every shade of constitutional difference, on the whole machinery of life. Man selects only for his own good: Nature only for the being which she tends. The slightest differences may turn the nicely balanced scale in the struggle for life, and so be preserved. How fleeting are the wishes and efforts of man! how short his time! and consequently how poor will be his results, compared with those accumulated by Nature during whole geological periods! Can we wonder, then, that Nature's productions should be far 'truer' in character than man's productions; that they should be infinitely better adapted to the most complex conditions of life, and should plainly bear the stamp of far higher workmanship?

It may metaphorically be said that natural selection is daily and hourly scrutinising, throughout the world, the slightest variations; rejecting those that are bad, preserving and adding up all that are good; silently and insensibly working, *whenever and wherever opportunity offers*, at the improvement of each organic being in relation to its conditions of life. We see nothing of these slow changes in progress, until the hand of time has marked the lapse of ages, and then so imperfect is our view into long-past geological ages, that we see only that the forms of life are now different from what they formerly were.

[In the next paragraph Darwin touches on one of the main problems which he believed his theory faced. He thought that a variation which arose in one generation would be diluted when passed on to the offspring, and he was forced to suppose that the same variation must arise again and again for a variety to be formed. Particulate inheritance, discovered by Gregor Mendel, allows the variation, in the form of genes, to be passed on to some
of the offspring without dilution (see diagram on p. 22). Theoretically a gene can be passed on for an infinite number of generations without any dilution of its effects.]

In order that any great modification should be effected in a species, a variety when once formed must again, perhaps after a long interval of time,

'No country can be named in which all the native inhabitants are so perfectly adapted that none of them could be still better adapted or improved: for in all countries the natives have allowed some foreigners to take firm possession of the land.' **The dog-like marsupials known as thylacines (above) were driven to extinction in Australia by the dingoes and are thought to survive only in a remote part of Tasmania. The dingoes are descended from domesticated dogs that were brought to Australia by the aborigines and then escaped into the wild. Dingoes excelled in hunting animals such as wallabies (below), and because of their competition the thylacines could not capture enough prey to survive.**

present individual differences of the same favour-able nature as before; and these must be again preserved, and so onwards step by step. Seeing that individual differences of the same kind per-petually recur, this can hardly be considered an unwarrantable assumption. But whether it is true, we can judge only by seeing how far the hypothesis accords with and explains the general phenomena of nature.

Although natural selection can act only through and for the good of each being, yet characters which we are apt to consider of trifling import-ance may thus be acted on. When we see leaf-eating insects green, and bark-feeders mottled-grey; the alpine ptarmigan white in winter, the red grouse the colour of heather, we must believe that these tints are of service to these birds and insects in preserving them from danger. Grouse suffer largely from hawks which are guided by eyesight to their prey. Hence natural selection gives the proper colour to each kind of grouse.

We have seen how the colour of the hogs which feed on the 'paint-root' in Virginia determines whether they shall live or die. In plants, the down on the fruit and the colour of the flesh are con-sidered by botanists of trifling importance: yet in the United States, cultivated smooth-skinned fruits suffer far more from a beetle than those with down; purple plums suffer far more from a certain disease than yellow plums; whereas another disease attacks yellow-fleshed peaches far more than those with other coloured flesh. In a state of nature, such differences would effectually settle which variety should succeed. In looking at small points of difference between species which seem quite unimportant, it is necessary to bear in mind that, owing to the law of correlation, when one part varies, other modifications, often of the most unexpected nature, will ensue.

As those variations which, under domestication, appear at any particular period of life, tend to reappear in the offspring at the same period – for instance in the shape, size, and flavour of seeds; in the caterpillar and cocoon stages of the varieties of the silkworm – so in nature, natural selection will be enabled to modify organic beings at any age, by the accumulation of variations profitable at that age. If it profit a plant to have its seeds more and more widely disseminated by the wind, I can see no greater difficulty in this being effected through natural selection, than in the cotton-planter increasing and improving by selection the down in the pods on his cotton-trees. Natural selection may adapt the larva of an insect to a score of con-tingencies, wholly different from those which concern the mature insect; and these modifications may affect, through correlation, the structure of the adult. So, conversely, modifications in the adult may affect the structure of the larva

Natural selection will modify the structure of the young in relation to the parent, and of the parent in relation to the young. A structure used only once in an animal's life, if of high importance to it, might be modified to any extent by natural selection; for instance, the hard tip to the beak of unhatched birds, used for breaking the egg. Of the best short-beaked tumbler-pigeons a greater number perish in the egg than are able to get out, so fanciers assist in the hatching. Now if nature had to make the beak of a full-grown pigeon very short for the bird's own advantage, the process of modification would be very slow, and there would be simultaneously the most rigorous selection of all the young birds within the egg which had the most powerful and hardest beaks, for all with weak beaks would inevitably perish: or more easily broken shells might be selected.

With all beings there must be much fortuitous destruction, which can have little or no influence on the course of natural selection. A vast number of eggs or seeds are annually devoured, and these could be modified through natural selection only if they varied in some manner which protected them from their enemies. Yet many of these eggs or seeds would perhaps, if not destroyed, have yielded better adapted individuals than any of those

'Sexual Selection . . . depends not on a struggle for existence in relation to other organic beings or external conditions, but on a struggle between the individuals of one sex, generally the males, for the possession of the other sex. The result is not death to the unsuccessful competitor, but few or no offspring.' Sexual selection is graphically illustrated by frogs in crowded conditions, when several males will grapple desperately to mate with a single female. In the picture above the female is in the centre, the male on the left is mating with her, while a second male clasps her from the other side in an unsuccessful attempt to mate. In most animals sexual selection is rather more subtle, involving ritualized disputes between males and complex courtship behaviours.

Male stag beetles fighting for possession of a female. The loser will not be killed, but will be unable to mate unless he wins another contest. Only the males have the enormous mandibles which give the species its name.

which happened to survive. So again a vast number of mature animals and plants must be annually destroyed by accidental causes, which would not be mitigated by changes of structure or constitution which would in other ways be beneficial to the species. But let the destruction be ever so heavy, if the number which can exist in any district be not wholly kept down, yet of those which do survive, the best adapted individuals will tend to propagate their kind in larger numbers than the less well adapted. If the numbers be wholly kept down, as will often have been the case, natural selection will be powerless in certain beneficial directions; but this is no valid objection to its efficiency at other times.

Sexual Selection

Inasmuch as peculiarities often appear under domestication in one sex and become hereditarily attached to that sex, so no doubt it will be under nature. Thus the two sexes can be modified through natural selection in relation to different habits of life, as is sometimes the case; or one sex can be modified in relation to the other sex, as commonly occurs. This leads me to say a few words on what I have called Sexual Selection.

This form of selection depends not on a struggle for existence in relation to other organic beings or external conditions, but on a struggle between the individuals of one sex, generally the males, for the possession of the other sex. The result is not death to the unsuccessful competitor, but few or no offspring. Sexual selection is, therefore, less rigorous than natural selection. Generally, the most vigorous males, those which are best fitted for their places in nature, will leave most progeny. But in many cases victory depends not so much on general vigour as on having special weapons confined to the male sex. A hornless stag or spurless cock would have a poor chance of leaving numerous offspring. Sexual selection, by always allowing the victor to breed, might surely give indomitable courage, length to the spur, and strength to the

'Amongst birds, the contest is often of a more peaceful character.' **To attract a mate the male frigate bird (*right*) inflates his bright red chest into a monstrous balloon; the female (*left*) has no red chest. Frigate birds nest on oceanic islands throughout the tropics: they are shown here on the Galapagos Islands.**

wing to strike with the spurred leg, in nearly the same manner as does the brutal cockfighter by the careful selection of his best cocks.

How low in the scale of nature the law of battle descends, I know not; male alligators have been described as fighting, bellowing and whirling round like Indians in a war-dance for the possession of the females; male salmon have been observed fighting all day long, male stag-beetles sometimes bear wounds from the huge mandibles of other males; the males of certain hymenopterous insects have been frequently seen by that inimitable observer M. Fabre fighting for a particular female, who sits by, an apparently unconcerned beholder of the struggle, and then retires with the conqueror. The war is, perhaps, severest between the males of polygamous animals, and these seem oftenest provided with special weapons. The males of carnivorous animals are already well armed; though to them and to others special

means of defence may be given through means of sexual selection, as the mane to the lion, and the hooked jaw to the male salmon; for the shield may be as important for victory as the sword or spear.

Amongst birds, the contest is often of a more peaceful character. There is the severest rivalry between the males of many species to attract, by singing, the females. The rock-thrush of Guiana, Birds-of-paradise, and some others congregate and successive males display with the most elaborate care and show off in the best manner their gorgeous plumage; they likewise perform strange antics before the females, which, standing by as spectators, at last choose the most attractive partner. Those who have closely attended to birds in confinement well know that they often take individual preferences and dislikes: thus Sir R. Heron has described how a pied peacock was eminently attractive to all his hen birds. If man can in a short time give beauty and an elegant carriage to his bantams, according to his standard of beauty, I see no good reason to doubt that female birds, by selecting, during thousands of generations, the most melodious or beautiful males, according to their standard of beauty, might produce a marked effect. Some well-known laws, with respect to the plumage of male and female birds, in comparison with the plumage of the young, can partly be explained through the action of sexual selection on variations occurring at different ages, and transmitted to the males alone or to both sexes at corresponding ages.

Thus, when the males and females of any animal have the same general habits of life, but differ in structure, colour, or ornament, such differences have been mainly caused by sexual selection: that is, by individual males having had, in successive generations, some slight advantage over other males, in their weapons, means of defence, or charms, which they have transmitted to their male offspring alone. Yet I would not wish to attribute all sexual differences to this agency: for we see in our domestic animals peculiarities

'Thus when the males and females of any animal have the same general habits of life, but differ in structure, colour, or ornament such differences have been mainly caused by sexual selection . . .' When John Gould painted this hummingbird, Loddiges' racquet-tail, he did not include the females since they are so different from the males that they were thought to be a different species. The females are larger, and their plumage less colourful. They lack the racquet-like tail feathers, which in the male are slapped sharply together to make a loud sound during disputes with other males and courtship flights.

arising and becoming attached to the male sex which apparently have not been augmented through selection by man. The tuft of hair on the breast of the wild turkey-cock cannot be of any use, and it is doubtful whether it can be ornamental in the eyes of the female bird; indeed, had the tuft appeared under domestication, it would have been called a monstrosity.

Illustrations of the Action of Natural Selection, or the Survival of the Fittest

Let us take the case of a wolf, which preys on various animals, securing some by craft, some by strength, and some by fleetness; and let us suppose that the fleetest prey, a deer, had from any change increased in numbers, or that other prey had decreased in numbers, during that season when the wolf was hardest pressed for food. Under such circumstances the swiftest and slimmest wolves would have the best chance of surviving, and so be preserved – provided always that they retained strength to master their prey. I may add that there are two varieties of wolf inhabiting the Catskill Mountains in the United States, one with a light greyhound-like form, which pursues deer, and the other more bulky, with shorter legs, which more frequently attacks the shepherd's flocks.

It may be worth while to give more complex illustration of the action of natural selection. Certain plants excrete sweet juice, apparently for the sake of eliminating something injurious from the sap. This juice is greedily sought by insects; but their visits do not in any way benefit the plant. Now, let us suppose that the juice or nectar was excreted from the inside of the flowers of a certain number of plants of any species. Insects in seeking the nectar would get dusted with pollen, and would often transport it from one flower to another. The flowers of two distinct individuals of the same species would thus get crossed; and the act of crossing, as can be fully proved, gives rise to vigorous seedlings, which would have the best chance of surviving. The plants which produced flowers with the largest glands or nectaries, excreting most nectar, would oftenest be visited by insects, and would oftenest be crossed; and so in the long-run would form a local variety. The flowers, also, which had their stamens and pistils placed in relation to the size and habits of the particular insect which visited them, so as to favour the transport of the pollen, would likewise be favoured.

When our plant had been rendered highly attractive to insects, they would, unintentionally, regularly carry pollen from flower to flower. Another process might now commence. No naturalist doubts the advantage of the 'physiological division of labour'; hence it would be advantageous to a plant to produce stamens alone in one flower or on one whole plant, and pistils

Darwin was mistaken in thinking that the tuft of hair on the breast of the wild turkey cock could not be the product of sexual selection. Any feature, however ugly it might seem to man, can become the focus of sexual selection and thus become exaggerated in size.

alone in another flower or on another plant. Now if this were to occur in ever so slight a degree under nature, individuals with this tendency would be continually favoured or selected, until at last a complete separation of the sexes might be effected.

Let us now turn to the nectar-feeding insects; we may suppose our plant to be a common plant, and certain insects to depend in main part on its nectar for food. I could give many facts showing how anxious bees are to save time – for instance, their habit of cutting holes and sucking the nectar at the bases of certain flowers, which with very little trouble they can enter by the mouth. Thus under certain circumstances individual differences in the curvature or length of the proboscis might profit a bee or other insect, so that certain individuals would be able to obtain their food more quickly than others; and thus the communities to which they belonged would throw off many swarms inheriting the same peculiarities.

The tubes of the corolla of the common red and incarnate [crimson] clovers do not on a hasty glance appear to differ in length; yet the hive-bee can easily suck the nectar out of the incarnate clover, but not out of the common red clover, which is visited by humble-bees [bumble-bees] alone. That this nectar is much liked by the hive-bee is certain; for I have repeatedly seen, but only in the autumn, many hive-bees sucking the flowers through holes bitten in the base of the tube by humble-bees. The difference in the length of the corolla in the two kinds of clover must be very trifling; for when red clover has been mown, the flowers of the second crop are somewhat smaller, and these are visited by many hive-bees. Thus, where this kind of clover abounded, it might be a great advantage to the hive-bee to have a slightly longer or differently constructed proboscis. On the other hand, as the fertility of this clover absolutely depends on bees visiting the flowers, if humble-bees were to become rare in any country, it might be a great advantage to the plant to have a shorter corolla, so that hive-bees should be enabled to suck its flowers. Thus a flower and a bee might slowly become, either simultaneously or one after the other, modified and adapted to each other in the most perfect manner by the continued preservation of all the individuals which presented slight deviations of structure mutually favourable to each other.

On the Intercrossing of Individuals

In the case of animals and plants with separated sexes, it is of course obvious that two individuals must always (with the exception of the curious cases of parthenogenesis) unite for each birth; but in the case of hermaphrodites this is far from obvious. Nevertheless there is reason to believe that with all hermaphrodites two individuals, either occasionally or habitually, concur for the reproduction of their kind.

With animals and plants a cross between different varieties, or between individuals of the same variety but of another strain, gives vigour and fertility to the offspring; on the other hand *close* interbreeding diminishes vigour and fertility. These facts alone incline me to believe that it is a general law of nature that no organic being fertilises itself for a perpetuity of generations; but that a cross with another individual is occasionally indispensable.

Exposure to wet is unfavourable to the fertilisation of a flower, yet what a multitude of flowers have their anthers and stigmas fully exposed to the weather! If an occasional cross be indispensable, the fullest freedom for the entrance of pollen from another individual will explain the exposure of the organs. Many flowers, on the other hand, have their organs closely enclosed, as in the papilionaceous or pea-family; but these show adaptations in relation to the visits of insects. So necessary are the visits of bees to many papilionaceous flowers, that their fertility is greatly diminished if these visits be prevented.

Now, it is scarcely possible for insects to fly from flower to flower, and not to carry pollen from one to the other. But bees will not thus produce a multitude of hybrids between distinct species; for if a plant's own pollen and that from another species are placed on the same stigma, the former invariably and completely destroys the influence of the foreign pollen. In numerous cases, far from self-fertilisation being favoured, there are special contrivances which effectually prevent the stigma receiving pollen from its own flower. In very many other cases, either the anthers burst before the stigma is ready for fertilisation, or the stigma is ready before the pollen of that flower is ready, so that these plants have in fact separated sexes, and must habitually be crossed.

If several varieties of the cabbage be allowed to

seed near each other, a large majority of the seed-lings turn out mongrels. Yet the pistil of each cabbage-flower is surrounded not only by its own six stamens, but by those of the many other flowers on the same plant; and the pollen of each flower readily gets on its own stigma without insect-agency. Such a vast number of mongrels must arise from the pollen of a distinct *variety* having a prepotent effect over the flower's own pollen. When distinct *species* are crossed the case is reversed, for a plant's own pollen is almost always pre-potent over foreign pollen.

In the case of a large tree it may be objected that pollen could seldom be carried from tree to tree, but trees have a strong tendency to bear flowers with separate sexes. When the sexes are separated, although the male and female flowers may be produced on the same tree, pollen must be regularly carried from flower to flower; and this will give a better chance of pollen being occasionally carried from tree to tree. That trees belonging to all Orders have their sexes more often separated than other plants, I find to be the case in this country, New Zealand, and the United States. On the other hand the rule does not hold good in Australia.

Various terrestrial animals are hermaphrodites, such as the land-mollusca [snails and slugs] and earth-worms; but these all pair. As yet I have not found a single terrestrial animal which can fer-tilise itself. This remarkable fact is intelligible on the view of an occasional cross being indispensable. Of aquatic animals, there are many self-fertilising hermaphrodites; but here the currents of water offer an obvious means for an occasional cross.

Circumstances favourable for the production of new forms through Natural Selection

This is an extremely intricate subject. A great amount of variability will evidently be favourable, but a large number of individuals will compensate for a lesser amount of variability and is a highly important element of success.

Isolation is also an important element in modi-fication through natural selection. In a confined or isolated area, if not very large, the organic and inorganic conditions of life will generally be almost uniform; so that natural selection will tend to modify all the varying individuals of the same species in the same manner. Intercrossing with inhabitants of the surrounding districts will be prevented. But even within the same area two varieties of the same animal may long remain distinct, from haunting different stations, from breeding at slightly different seasons, or from the individuals of each variety preferring to pair together.

Isolation likewise prevents, after any physical change, as of climate, elevation, etc., the immigra-tion of better adapted organisms; and thus new places in the natural economy will be filled up by

'Various terrestrial animals are hermaphrodites, such as the land-mollusca and the earth-worms; but all these pair. As yet I have not found a single terrestrial animal which can fertilize itself.'
Terrestrial hermaphrodites, such as these snails, always mate before reproduction despite the fact that every individual has both male and female organs. The advantage of one snail exchanging sperm with another is that it increases variability among the offspring.

modification of the old inhabitants. Lastly, isolation will give time for a new variety to be improved at a slow rate. If, however, an isolated area be very small, the total number of inhabitants will be small, and this will retard the production of new species, by decreasing the chances of favourable variations arising.

The mere lapse of time by itself does nothing, either for or against natural selection. It has been erroneously asserted that the element of time has been assumed by me to play an all-important part in modifying species, as if all the forms of life were necessarily undergoing change through some innate law. Lapse of time is only so far important, that it gives a better chance of beneficial variations arising, and of their being selected, accumulated, and fixed.

If we look at any small isolated area, such as an oceanic island, although the number of species inhabiting it is small, yet of these species a very large proportion are endemic – that is, have been produced there, and nowhere else in the world. Hence an oceanic island seems to have been highly favourable for the production of new species. But we may thus deceive ourselves, for to ascertain whether a small isolated area, or a continent, has been most favourable, we ought to make the comparison within equal times; and this we are incapable of doing.

On the whole I believe that largeness of area is more important than isolation, especially for the production of species which shall endure for a long period and spread widely. Throughout a great and open area not only will there be a better chance of favourable variations arising from the large number of individuals of the same species there supported, but the conditions of life are much more complex from the large number of species; and if some species become improved, others will have to be improved in a corresponding degree, or they will be exterminated. Each improved form will be able to spread over the open area, and will thus come into competition with many other forms. Moreover, great areas, though now continuous, will often, owing to former oscillations of level, have existed in a broken condition as islands; so the good effects of isolation will, to a certain extent, have concurred. Finally, modification will generally have been more rapid on large areas; and the new forms produced on large areas, which already have been victorious over many competitors, will spread most widely and give rise to the greatest number of new varieties and species. They will thus play a more important part in the organic world.

In accordance with this view, we can understand the fact of the productions of the smaller continent of Australia yielding before those of the larger Europaeo-Asiatic area, and that continental productions have everywhere become so largely naturalised on islands. On a small island the race for life will have been less severe, and there will have been less modification and extermination. Hence the flora of Madeira resembles to a certain extent the extinct Tertiary flora of Europe. All freshwater basins make a small area compared with that of the sea or land. Consequently, the competition will have been less severe and old forms more slowly exterminated. And it is in fresh-water basins that we find seven genera of Ganoid fishes, remnants of a once preponderant order, and the Ornithorhynchus [duck-billed platypus] and Lepidosiren [a lungfish], which, like fossils, connect to a certain extent orders at present widely sundered in the natural scale. These anomalous forms may be called living fossils; they have endured from having been exposed to less varied, and therefore less severe, competition.

Divergence of Character
This principle is of high importance, and explains how the lesser difference between varieties becomes augmented into the greater difference between species. Mere chance might cause one variety to differ in some character from its parents, and the offspring of this variety again to differ in a

greater degree; but this alone would never account for so habitual and large a degree of difference as that between species.

As has always been my practice, I have sought light from our domestic productions. It will be admitted that the production of races so different as Short-horn and Hereford cattle, the several breeds of pigeons, etc., could never have been effected by the mere chance accumulation of similar variations during many generations. In practice, a fancier is, for instance, struck by a pigeon having a slightly shorter beak; another fancier is struck by a pigeon having a rather longer beak; and on the principle that 'fanciers will not admire a medium standard, but like extremes', they both go on choosing and breeding from birds with longer and longer beaks, or with shorter and shorter beaks. Again, we may suppose that at an early period of history, the men of one nation required swifter horses, whilst those of another required stronger horses. In time, from the continued selection of swifter horses in the one case, and of stronger ones in the other, the differences would become greater and form two sub-breeds. After centuries these sub-breeds would become two distinct breeds. As the differences became greater, the inferior animals with intermediate characters, being neither very swift nor very strong, would not have been used for breeding, and will thus have tended to disappear. Here, then, the principle of divergence causes differences steadily to increase, and the breeds to diverge in character both from each other and from their common parent.

But how can any analogous principle apply in nature? I believe it can, from the simple circumstance that the more diversified the descendants from any species become, by so much will they be better enabled to seize on diversified places in nature, and so increase in numbers.

Take the case of a carnivorous quadruped, of which the number that can be supported in any country has long ago arrived at its full average. If its natural power of increase be allowed to act, it can succeed in increasing (the country not undergoing any change in conditions) only by its varying descendants seizing on places at present occupied by other animals: some of them feeding on new kinds of prey; some inhabiting trees, frequenting water, and some perhaps becoming less carnivorous. The more diversified the descendants become, the more places they will be enabled to occupy.

So it will be with plants. If a plot of ground be sown with one species of grass, and a similar plot be sown with several distinct genera of grasses, a greater number of plants and a greater dry weight can be raised in the latter case. Hence, if the varieties of any one species of grass were continually selected which differed from each other in a slight degree, a greater number of individual plants of this species, including its modified descendants, would succeed in living on the same piece of ground.

That the greatest amount of life can be supported by diversification of structure is seen under many circumstances. Farmers find that they can raise most food by a rotation of plants belonging to the most different orders: nature follows what may be called a simultaneous rotation. The advantages of diversification of structure, with the accompanying differences of habit and constitution, determine that the inhabitants on any small piece of ground shall, as a general rule, belong to what we call different genera and orders.

The same principle is seen in the naturalisation of plants through man's agency in foreign lands. It might have been expected that the plants which would succeed in becoming naturalised would have been closely allied to the indigenes; for these are commonly looked at as specially created for their own country. But in the last edition of Dr. Asa Gray's *Manual of the Flora of the Northern United States*, 260 naturalised plants are enumerated, belonging to 162 genera; and out of these 162 naturalised genera, no less than 100 genera are not there indigenous; thus a large proportional addition has been made to the genera living in the United States.

The advantage of diversification in the inhabitants of the same region is the same as that of the physiological division of labour in the organs of the individual body. A stomach adapted to digest vegetable matter alone, or flesh alone, draws most nutriment from these substances. So in the general economy of any land, the more widely and perfectly animals and plants are diversified for different habits of life, so will a greater number of individuals be capable of supporting themselves.

The Probable Effects of the Action of Natural Selection through Divergence of Character and Extinction, on the Descendants of a Common Ancestor

After the foregoing discussion, we may assume that the modified descendants of any one species will succeed better as they become more diversified in structure, and are thus enabled to encroach on places occupied by other beings. Now let us see how this principle, combined with the principles of natural selection and extinction, tends to act.

The diagram below will aid us in understanding this rather perplexing subject. Let A to L represent the species of a genus; these species are supposed to resemble each other in unequal degrees, as represented in the diagram by the letters standing at unequal distances. Let A be a common, widely diffused, and varying species. The diverging dotted lines proceeding from A, represent its varying offspring. Here the principle of benefit derived from divergence of character comes in, for this will generally lead to the most divergent variations (represented by the outer dotted lines) being preserved and accumulated by natural selection. When a dotted line reaches one of the horizontal lines, a sufficient amount of variation has been accumulated to form it into a fairly well-marked variety.

The intervals between the horizontal lines in the diagram may represent each a thousand or more generations. After a thousand generations, species A has produced two fairly well-marked varieties, namely a^1 and m^1. These two varieties being only slightly modified forms, will tend to inherit those advantages which made their parent A more numerous than most of the other inhabitants of the same country. If then, these two varieties be variable, the most divergent of their variations will generally be preserved during the next thousand generations. And after this interval, variety a^1 has produced variety a^2, which will, owing to the principle of divergence, differ more from A than did variety a^1. Variety m^1 is supposed to have produced two varieties, namely m^2 and s^2, differing from each other, and more considerably from their common parent, A.

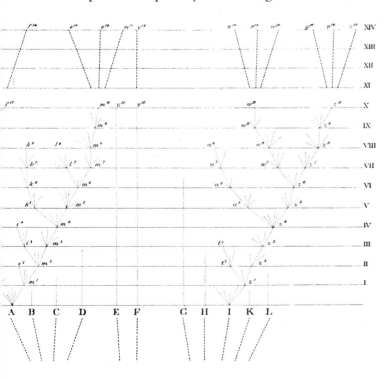

Darwin's diagram of a hypothetical evolutionary tree. The letters A–L represent eleven species of a genus. The horizontal lines represent intervals of about a thousand generations. The dotted lines trace the descendants of each species, and the small letters denote *'well-marked varieties'*. For further explanation see the text.

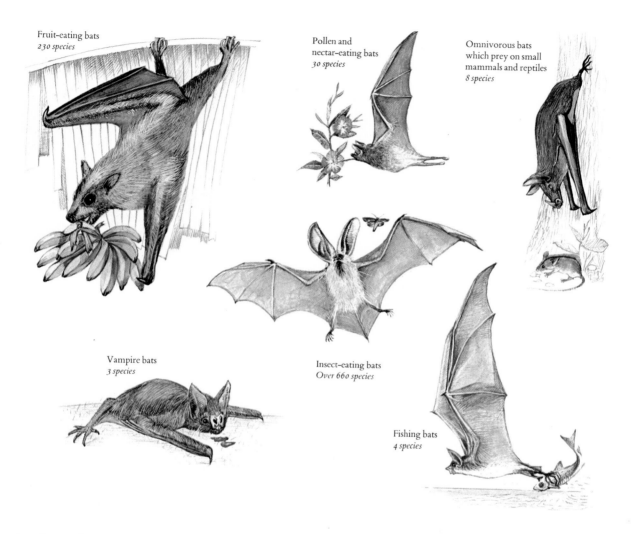

Fruit-eating bats
230 species

Pollen and
nectar-eating bats
30 species

Omnivorous bats
which prey on small
mammals and reptiles
8 species

Vampire bats
3 species

Insect-eating bats
Over 660 species

Fishing bats
4 species

'. . . the more diversified in structure the descendants from any one species, the more places they will be enabled to seize on and the more their modified progeny will increase.' **The beginnings of such a process of diversification can be observed in bats. There are two major groups of bats, the Megachiroptera which are basically fruit eaters, and the Microchiroptera which are basically insect eaters. Among the Megachiroptera some forms have evolved that feed mainly on pollen and nectar. Among the Microchiroptera far greater diversification has taken place, and forms have evolved that are specialized for feeding on fish, on blood, on fruit, and on pollen and nectar. Some have a varied diet taking fruit, nectar and insects for example. A few supplement this diet with small mammals and reptiles. This diversification in feeding involves relatively few species by comparison with the enormous number of insect-eating Microchiroptera, but it is probable that we are observing only the very early beginnings of diversification from which a substantial increase in the number of species could follow.**

I must here remark that I do not suppose that the process ever goes on so regularly as is represented in the diagram, nor that it goes on continuously; it is far more probable that each form remains for long periods unaltered, and then again undergoes modification. Nor do I suppose that the most divergent varieties are invariably preserved: a medium form may often long endure, for natural selection will always act according to the nature of the places which are not perfectly occupied by other beings, and this will depend on infinitely complex relations. But as a general rule, the more diversified in structure the descendants from any one species, the more places they will be enabled to seize on, and the more their modified progeny will increase. In our diagram the line of succession is broken at regular intervals by small numbered letters marking the successive forms which have become sufficiently distinct to be recorded as varieties. But these breaks are imaginary, and might have been inserted anywhere, after intervals long enough to allow the accumulation of a considerable amount of divergent variation.

The modified offspring from the later and more highly improved branches in the lines of descent will often take the place of the earlier and less improved branches: this is represented in the diagram by some of the lower branches not reaching to the upper horizontal lines.

After ten thousand generations, species A has produced a^{10}, f^{10}, and m^{10}, which will have come to differ largely, but unequally, from each other and their common parent. These three forms may still be only well-marked varieties; but we have only to suppose the steps of modification to be more numerous or greater in amount, to convert these three forms into species. Thus the diagram illustrates how the small differences distinguishing varieties are increased into the larger differences distinguishing species. By continuing the process for a greater number of generations, (as shown in the diagram in a condensed and simplified manner), we get eight species, marked by the letters between a^{14} and m^{14}, all descended from A. Thus species are multiplied and genera formed.

In a large genus it is probable that more than one species would vary. In the diagram a second species, I, has produced, after ten thousand generations, either two varieties (w^{10} and z^{10}) or two species, according to the amount of change represented between the horizontal lines. After fourteen thousand generations, six new species, marked by the letters n^{14} to z^{14}, have been produced. In any genus the species already very different from each other will generally produce the greatest number of modified descendants; for these will have the best chance of seizing new and different places in nature: hence in the diagram the species A and I have given rise to new varieties and species. The other species of our original genus may for long but unequal periods transmit unaltered descendants; this is shown by the dotted lines prolonged upwards.

But another principle, namely extinction, plays an important part. As in each fully stocked country natural selection necessarily acts by the selected form having some advantage in the struggle for life over other forms, the improved descendants of any one species tend to supplant and exterminate their predecessors and their progenitor. For the competition will generally be most severe between those forms most nearly related in habits and structure. Hence all the intermediate forms between the earlier and later states, as well as the parent-species, tend to become extinct. So it will be with whole collateral lines of descent, which will be conquered by later and improved lines. If, however, modified offspring get into some distinct country or quickly adapt to some quite new station in which offspring and progenitor do not come into competition, both may continue to exist.

If, then, our diagram represents considerable modification, species A and all the earlier varieties will have become extinct, being replaced by eight new species, (a^{14} to m^{14}); and species I will be replaced by six new species, (n^{14} to z^{14}).

These two species, A and I, were widely diffused, having some advantage over most of the other species of the genus. Their descendants at the fourteen-thousandth generation will have inherited some of the same advantages: they have also been modified and improved in a diversified manner. Therefore, probably they will have taken the places of some of the original species most nearly related to their parents. Hence very few of the original species will have transmitted offspring to the fourteen-thousandth generation.

The new species in our diagram descended

from the original eleven species will now be fifteen in number. Owing to divergence, the difference between a^{14} and z^{14} will be much greater than that between the most distinct of the original species. The new species, moreover, will be allied in a widely different manner. Of the eight descendants from A, the three marked a^{14}, q^{14}, p^{14} will be nearly related having recently branched off from a^{10}; b^{14} and f^{14}, from having diverged at an earlier period from a^5, will be in some degree distinct from these three; o^{14}, e^{14}, and m^{14}, having diverged at the commencement of modification, will be widely different from the other five species, and may constitute a sub-genus or distinct genus.

The six descendants from I will form two sub-genera or genera. But as the original species I differed largely from A, the two groups have diverged in different directions. The intermediate species, which connected the original species A and I, have all become, excepting F, extinct and have left no descendants. Hence the six species descended from I and the eight from A have to be ranked as very distinct genera, or even sub-families.

Thus two or more genera are produced, by descent with modification, from two or more species of the same genus. And the two or more parent-species are descended from some one species of an earlier genus. In our diagram this is indicated by the broken lines beneath the capital letters, converging downwards towards a single point.

The new species F^{14} has curious affinities to the other fourteen new species. Descended from a form which stood between the parent-species

'For the competition will generally be most severe between those forms most nearly related in habits and structure. Hence all the intermediate forms between the earlier and later states, as well as the parent species, tend to become extinct.' This general principle of biological evolution is well illustrated by the evolution of *Homo sapiens* (*front*). The successive species of *Homo* have replaced the earlier species, their ape-like ancestor *Ramapithecus* (*top right*), and the related line of *Australopithecus* (*left*).

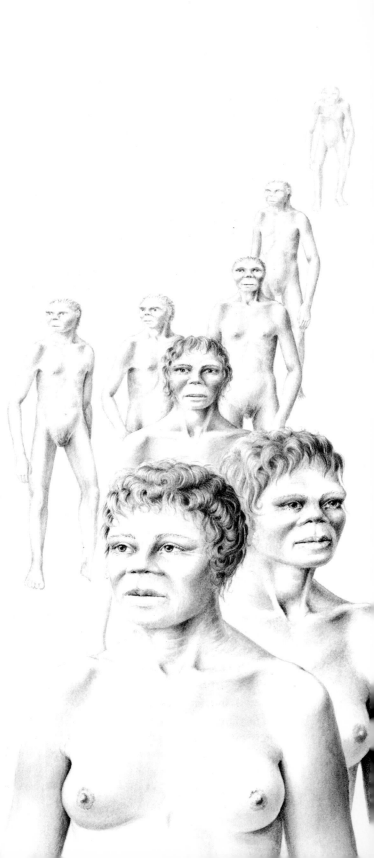

A and I, now extinct and unknown, it is in some degree intermediate between the two groups descended from these species. But as these two groups have gone on diverging from their parents, the new species F^{14} will not be directly intermediate between them, but rather between types of the two groups.

If, in the diagram, the amount of change represented by each successive group of diverging dotted lines is great, the forms a^{14} to p^{14}, those marked b^{14} and f^{14}, and those marked o^{14} to m^{14} will form three very distinct genera. We shall also have two very distinct genera descended from I, differing widely from the descendants of A. These two groups of genera will form two distinct families, or orders, according to the amount of modification supposed to be represented in the diagram.

It is the species belonging to the larger genera which oftenest present varieties. Hence, the struggle for the production of modified descendants will mainly lie between the larger groups. One large group will slowly conquer another, reduce its numbers, and thus lessen its chance of further variation and improvement. Within the same large group, the later and more highly perfected sub-groups tend to supplant the earlier and less improved sub-groups. Small and broken groups and sub-groups will finally disappear. The groups of organic beings which are now large and triumphant will, for a long period, continue to increase. But which groups will ultimately prevail, no man can predict. Owing to the continued increase of the larger groups, a multitude of smaller groups will become extinct; consequently, of the species now living, extremely few will transmit descendants to a remote futurity. Accordingly, extremely few ancient species have transmitted descendants to the present day, and, as all the descendants of the same species form a class, we can understand how it is that so few classes exist in each main division of the animal and vegetable kingdoms.

On the Degree to which Organisation tends to advance

Natural Selection acts exclusively by the preservation and accumulation of variations which are beneficial under the conditions to which each creature is exposed. The ultimate result is that each creature tends to become more and more improved in relation to its conditions. This improvement leads to the gradual advancement of the organisation of the greater number of living beings throughout the world. But naturalists have not defined to each other's satisfaction what is meant by an advance in organisation. Amongst the Vertebrata the degree of intellect and an approach in structure to man clearly come into play. It might be thought that the amount of change which the various organs pass through in their development from the embryo to maturity would suffice as a standard of comparison; but there are cases, as with certain parasitic crustaceans, in which several parts of the structure become less perfect, so that the mature animal cannot be called higher than its larva. Von Baer's standard seems the most widely applicable and the best, namely, the amount of differentiation of the parts of the same organic being, and their specialisation for different functions; or the completeness of the division of physiological labour.

Natural selection clearly leads towards this standard: all physiologists admit that the specialisation of organs, inasmuch as in this state they perform their functions better, is an advantage. On the other hand, natural selection can gradually fit a being to a situation in which several organs would be superfluous: in such cases there would be retrogression in the scale of organisation.

But if all organic beings thus tend to rise in the scale, how is it that a multitude of the lowest forms still exist; and that in each great class some forms are far more highly developed than others? Why have not the more highly developed forms everywhere supplanted and exterminated the lower? Lamarck, who believed in an innate tendency

Springtails are primitive wingless insects. None of their ancestors had wings.

Dragonflies have wings and therefore represent an evolutionary advance over springtails. The two pairs of wings are identical.

Beetles represent a further advance since the front pair of wings are specialized as hard covers (elytra) which protect the wings and body.

A flightless beetle from Madeira. It has clearly evolved from a flying beetle but the trend of greater differentiation and specialization has been reversed.

Natural selection tends to produce greater differentiation and specialization, or what Darwin calls *'an advance in organization',* **but it need not necessarily do so, and it can have the reverse effect, as with the evolution of flightless beetles on islands.**

towards perfection in all organic beings, seems to have felt this difficulty so strongly that he was led to suppose that new and simple forms are continually being produced by spontaneous generation. Science has not as yet proved the truth of this belief, whatever the future may reveal.

On our theory the continued existence of lowly organisms offers no difficulty; for natural selection or the survival of the fittest does not necessarily include progressive development – it only takes advantage of such variations as arise and are beneficial to each creature. And what advantage would it be to an infusorian animalcule, an intestinal worm, or even an earthworm to be highly organised? If it were no advantage these forms would be left unimproved or but little improved, and might remain for indefinite ages in their present lowly condition. But to suppose that most of the existing low forms have not advanced since the first dawn of life would be extremely rash; for every

naturalist who has dissected some of the beings now ranked as very low must have been struck with their wondrous and beautiful organisation.

Nearly the same remarks are applicable if we look to the different grades of organisation within the same group; for instance, in the vertebrata, to the co-existence of mammals and fish – amongst mammalia, to the co-existence of man and the Ornithorhynchus [duck-billed platypus] – amongst fishes, to the co-existence of the shark and the lancelet (*Amphioxus*), which latter fish in the extreme simplicity of its structure approaches the invertebrate classes. But the advancement of mammals to the highest grade would not lead to their taking the place of fishes. In some cases lowly organised forms have been preserved from inhabiting confined or peculiar stations, where they have been subjected to less severe competition, and where their scanty numbers have retarded the chance of favourable variations arising.

Looking to the first dawn of life, when all organic beings presented the simplest structure, how could the first steps in the differentiation of parts have arisen? As we have no facts to guide us, speculation on the subject is almost useless. It is, however, an error to suppose that there would be no struggle for existence and consequently no natural selection until many forms had been produced: variations in a single species might be beneficial, and thus the whole mass of individuals might be modified.

Indefinite Multiplication of Species

Mr. Watson has objected that the continued action of natural selection would tend to make an indefinite number of specific forms. There seems at first sight no limit to the amount of profitable diversification of structure, and therefore no limit to the number of species which might be produced. But geology shows us that from an early part of the Tertiary period the number of species of shells, and from the middle part the number of mammals, has scarcely increased. What then checks an indefinite increase in the number of species? The amount of life supported on an area must have a limit, depending as it does on physical conditions; therefore, if an area be inhabited by very many species, each will be represented by few individuals and will be liable to extermination from accidental fluctuations in the nature of the seasons or in the number of their enemies. The process of extermination in such cases would be rapid, whereas the production of new species must always be slow. Rare species, and each species will become rare if the number of species becomes indefinitely increased, will present few favourable variations; consequently, the process of giving birth to new specific forms would be retarded. When species become very rare, close interbreeding will help to exterminate them; this comes into play in accounting for the deterioration of the Aurochs in Lithuania and Bears in Norway. Lastly a dominant species, which has already beaten many competitors, tends to spread and supplant many others, thus checking the inordinate increase of specific forms throughout the world.

Summary of Chapter

If under changing conditions of life organic beings present individual differences, it would be a most extraordinary fact if no variations had ever occurred useful to each being's own welfare, in the same manner as so many variations have occurred useful to man. But if useful variations occur, individuals thus characterised will have the best chance of being preserved in the struggle for life; and from the strong principle of inheritance, these will tend to produce offspring similarly characterised. This principle of preservation, or the survival of the fittest, I have called Natural Selection. It leads to the improvement of each creature in relation to its conditions of life; and consequently, in most cases, to what must be regarded as an advance in organisation. Nevertheless, low and simple forms will long endure if well fitted for their simple conditions of life.

Natural selection, on the principle of qualities being inherited at corresponding ages, can modify the egg, seed or young as easily as the adult. Amongst many animals sexual selection will have given its aid to ordinary selection by assuring to the most vigorous and best adapted males the greatest number of offspring. Sexual selection will also give characters useful to the males alone, in their struggles or rivalry with other males; and these characters will be transmitted to one sex or to both sexes, according to the form of inheritance which prevails.

Whether natural selection has really thus acted in adapting the various forms of life to their several conditions and stations, must be judged by the general tenor and balance of evidence given in the following chapters. But we have already seen how it entails extinction; and how largely extinction has acted in the world's history, geology plainly declares. Natural selection, also, leads to divergence of character; for the more organic beings diverge in structure, habits, and constitution, by so much the more can a large number be supported on the same area. Thus the small differences distinguishing varieties of the same species steadily tend to increase, till they equal the greater differences between species of the same genus, or even of distinct genera.

It is a truly wonderful fact – the wonder of which we are apt to overlook from familiarity – that all animals and all plants throughout all time and space should be related to each other in groups subordinate to groups, in the manner which we everywhere behold – namely, varieties of the same species most closely related, species of the same genus less closely and unequally related, species of distinct genera much less closely related, and genera related in different degrees, forming sub-families, families, orders, sub-classes and classes. The several groups in any class cannot be ranked in a single file, but seem clustered round points, and these round other points. If species had been independently created, no explanation would have been possible of this kind of classification; but it is explained through inheritance and the complex action of natural selection, entailing extinction and divergence of character, as we have seen illustrated in the diagram.

The affinities of all the beings of the same class have sometimes been represented by a great tree. I believe this simile largely speaks the truth. The green and budding twigs may represent existing species; and those produced during former years may represent the long succession of extinct species. At each period of growth all the growing twigs have tried to branch out on all sides and to overtop and kill the surrounding twigs and branches, in the same manner as species and groups of species have at all times overmastered other species in the great battle for life. Of the many twigs which flourished when the tree was a mere bush, only two or three, now grown into great branches, yet survive and bear the other branches; so with the species which lived during long-past geological periods, very few have left living and modified descendants. From the first growth of the tree, many a limb and branch has decayed and dropped off; and these fallen branches of various sizes may represent those whole orders, families, and genera which have now no living representatives, and which are known to us only in a fossil state. As we here and there see a thin straggling branch springing from a fork low down in a tree, and which by some chance has been favoured and is still alive on its summit, so we occasionally see an animal which in some small degree connects by its affinities two large branches of life, and which has apparently been saved from fatal competition by having inhabited a protected station.

So it is that the great Tree of Life fills with its dead and broken branches the crust of the earth, and covers the surface with its ever-branching and beautiful ramifications.

Chapter 5
Laws of Variation

[*The variation upon which natural selection acts arises from gene recombination and mutation. Genetic variations are, in fact, 'due to chance', since the form they take is not influenced by the environmental conditions. Changes are produced in an individual's phenotype by the environment, but these cannot be passed on to the offspring. See also the notes on p. 50.*]

I have hitherto sometimes spoken as if variations were due to chance. This, of course, is a wholly incorrect expression, but it serves to acknowledge plainly our ignorance of the cause of variation. In the first chapter I attempted to show that changed conditions act in two ways, directly on the whole organisation or on certain parts alone, and indirectly through the reproductive system.

It is very difficult to decide how far changed conditions, such as of climate, food, etc., have acted. There is reason to believe that in the course of time the effects have been greater than can be clearly proved. But the innumerable complex co-adaptations of structure, which we see throughout nature, cannot be attributed simply to such action. In the following cases the conditions seem to have produced some slight definite effect: shells at their southern limit, and when living in shallow water, are often more brightly coloured than those of the same species from farther north or from a greater depth; birds of the same species are more brightly coloured under a clear atmosphere than when living near the coast or on islands; residence near the sea also affects the colours of insects. Certain plants, when growing near the sea-shore, have their leaves in some degree fleshy, though not elsewhere fleshy.

When a variation is of use to any being, we cannot tell how much to attribute to the accumulative action of natural selection, and how much to the conditions of life. Thus animals of the same species have thicker and better fur the farther north they live; but how much of this difference may be due to the warmest-clad individuals having been favoured and preserved during many generations, and how much to the action of the severe climate? For it would appear that climate has some direct action on the hair of our domestic quadrupeds. On the other hand innumerable instances are known of species keeping true although living under the most opposite climates. Such considerations incline me to lay less weight on the direct action of the surrounding conditions than on a tendency to vary, due to causes of which we are quite ignorant.

Effects of the increased Use and Disuse of Parts, as controlled by Natural Selection
[*The changes caused by use or disuse of an organ cannot be inherited.*]

I think there can be no doubt that use in our domestic animals has strengthened and enlarged certain parts, and disuse diminished them; and that such modifications are inherited. Under free nature many animals possess structures which can be best explained by the effects of disuse. It is probable that the nearly wingless condition of several birds which inhabit oceanic islands tenanted by no beast of prey, has been caused by disuse, as the larger ground-feeding birds seldom take flight except to escape danger. The ostrich indeed inhabits continents but it can defend itself by kicking its enemies.

Flightless cormorants on the Galapagos Islands. Darwin believed that they had lost their powers of flight through disuse. Since we now know that acquired characters are not inherited this cannot be true. The probable explanation is that once the wings were no longer in use, the energy that went into the development of the wings was being wasted, and any mutation that decreased their size would be advantageous since it would save energy.

But in some cases we might easily put down to disuse modifications which are wholly or mainly due to natural selection. Out of the 550 species of beetles inhabiting Madeira, 200 are so far deficient in wings that they cannot fly. Several facts – namely, that beetles in many parts of the world are frequently blown to sea and perish; that the beetles in Madeira lie much concealed until the wind lulls and the sun shines; that the proportion of wingless beetles is larger on the exposed Desertas than in Madeira itself; and especially that certain large groups of beetles, elsewhere excessively numerous, which absolutely require the use of their wings, are here almost entirely absent – make me believe that the wingless condition of so many Madeira beetles is mainly due to the action of natural selection, combined probably with disuse. For during many successive generations each beetle which flew least, either from its wings having been less perfectly developed or from indolent habit, will have had the best chance of surviving while those which most readily took to flight would oftenest have been blown to sea, and thus destroyed.

The eyes of moles and some burrowing rodents are rudimentary in size, and in some cases quite covered by skin and fur. This state of the eyes is probably due to gradual reduction from disuse, aided perhaps by natural selection. In South America a burrowing rodent, the tuco-tuco or *Ctenomys*, is even more subterranean in its habits than the mole, and is frequently blind. One which I kept alive was in this condition, the cause, as appeared on dissection, having been inflammation of the nictitating membrane. As frequent inflammation of the eyes must be injurious to any animal, and as eyes are certainly not necessary to animals having subterranean habits, a reduction in their size, with the adhesion of the eyelids and growth of fur over them, might in such case be an advantage; and natural selection would aid the effects of disuse.

It is well known that several animals, belonging to the most different classes, which inhabit the caves of Carniola and of Kentucky, are blind. In some of the crabs the foot-stalk for the eye remains, though the eye is gone. As it is difficult to imagine that eyes, though useless, could be in any way injurious to animals living in darkness, their loss may be attributed to disuse.

It is difficult to imagine conditions of life more similar than deep limestone caverns under a nearly similar climate; so that, in accordance with the old view of the blind animals having been separately created for the American and European caverns, close similarity in their organisation and affinities might have been expected. This is certainly not the case if we look at the two faunas.

On my view we must suppose that American animals slowly migrated by successive generations into the Kentucky caves, as did European animals into the caves of Europe. By the time an animal has reached the deepest recesses disuse will have more or less obliterated its eyes, and natural selection will have affected other changes, such as an increase in the length of the antennae or palpi, as a compensation for blindness. Notwithstanding such modifications, we still see in some cave-animals of America affinities to the other inhabitants of that continent, while some of the European cave-insects are very closely allied to those of the surrounding country. It would be difficult to give any rational explanation of the affinities of the blind cave-animals to the other inhabitants of the two continents on the ordinary view of their independent creation.

Another blind genus (*Anophthalmus*) offers this remarkable peculiarity, that the species have not been found anywhere except in caves; yet those which inhabit the caves of Europe and America are distinct; but it is possible that the progenitors of these several species, whilst they were furnished with eyes, may formerly have ranged over both

continents, and then have become extinct, excepting in their present secluded abodes. Far from feeling surprise that some of the cave-animals should be anomalous, I am only surprised that more wrecks of ancient life have not been preserved, owing to the less severe competition to which the inhabitants of these dark abodes will have been exposed.

Acclimatisation

As distinct species belonging to the same genus inhabit hot and cold countries, if it be true that all the species of the same genus are descended from a single parent-form, acclimatisation must be readily effected during a long course of descent. It is notorious that each species is adapted to the climate of its own home: species from an arctic or even from a temperate region cannot endure a tropical climate, or conversely. But the degree of adaptation of species to the climates under which they live is often overrated, and a number of plants and animals brought from different countries are here perfectly healthy. In regard to animals, several instances could be adduced of species having largely extended, within historical times, their range from warmer to cooler latitudes, and conversely.

The rat and mouse have been transported by man to many parts of the world, and now have a far wider range than any other rodent; for they live under the cold climate of Faroe in the north and of the Falklands in the south, and on many an island in the torrid zones. Adaptation to any special climate may be looked at as a quality readily grafted on an innate wide flexibility of constitution, common to most animals. The capacity of endur-

'The rat and mouse have been transported by man to many parts of the world, and now have a far wider range than any other rodent; for they live under the cold climate of Faroe in the north and of the Falklands in the south, and on many an island in the torrid zones.' The brown rat, *Rattus norvegicus*, was originally native to Asia, but is now found throughout the world, except in deserts and in the polar regions. It invaded Europe in the sixteenth century and from there spread to America in about 1775. This painting by J. J. Audubon shows brown rats devouring melons.

'*The capacity of enduring the most different climates by man and his domestic animals . . . ought not to be looked at as anomalies, but as examples of a very common flexibility of constitution, brought, under peculiar circumstances, into action.*' Man can, with time, adapt to living at high altitude where there is less oxygen. The main change that occurs is that the number of red blood cells increases, thus improving the efficiency of the blood in carrying oxygen from the lungs to the rest of the body. This adaptation to altitude is familiar to mountaineers, and is an example of the '*flexibility of constitution*' that Darwin mentions. Tribes who live permanently at high altitude, such as the Sherpas of the Himalayan mountains, show adaptations that are genetically determined, notably an increase in the size of their lungs.

ing the most different climates by man and his domestic animals, and the fact of the extinct elephant and rhinoceros having formerly endured a glacial climate, whereas the living species are now all tropical or sub-tropical in their habits, ought not to be looked at as anomalies, but as examples of a very common flexibility of constitution brought, under peculiar circumstances, into action.

How much of the acclimatisation of species to any peculiar climate is due to mere habit, and how much to the natural selection of varieties having different innate constitutions is an obscure question. I must believe that habit has some influence, but the effects have often been largely combined with, and sometimes overmastered by, natural selection.

Correlated Variation
[*See note on p. 51.*]
I mean by this expression that the whole organisation is so tied together during its growth and development, that when slight variations occur in one part and are accumulated through natural selection, other parts become modified. This is a very important subject, most imperfectly understood, and no doubt wholly different classes of facts may here be confounded together.

We may often falsely attribute to correlated variation structures which in truth are simply due to inheritance; for an ancient progenitor may have acquired through natural selection some one modification in structure, and, after thousands of generations, some other and independent modification; and these two modifications having been transmitted to a whole group of descendants would naturally be thought to be correlated.

Compensation and Economy of Growth

The law of compensation or balancement of growth holds true to a certain extent with our domestic productions: if nourishment flows to one part in excess, it rarely flows, at least in excess, to another part; thus it is difficult to get a cow to give much milk and to fatten readily. The same varieties of cabbage do not yield abundant foliage and a copious supply of oil-bearing seeds. When the seeds in our fruits become atrophied, the fruit itself gains largely in size and quality. With species in a state of nature the law is hardly of universal application; but many good observers, especially botanists, believe in its truth.

I suspect that as a more general principle natural selection is continually trying to economise every part of the organisation. If under changed conditions a structure becomes less useful, its diminution will be favoured, for it will profit the individual not to waste its nutriment in building up a useless structure. Thus when a cirripede is parasitic within another cirripede and is thus protected, it loses more or less completely its own shell.

Thus natural selection will tend to reduce any superfluous part of the organisation, without by any means causing some other part to be largely developed. And conversely natural selection may develop an organ without reducing some adjoining part.

Multiple, Rudimentary, and Lowly-organised Structures are Variable

It seems to be a rule, as remarked by Is. Geoffroy St. Hilaire, both with varieties and species, that when any part or organ is repeated many times in the same individual (as the vertebrae in snakes, and the stamens in polyandrous flowers) the number is variable; whereas the same part or organ, when it occurs in lesser numbers, is constant. Beings which stand low in the scale of nature are more variable than those which are higher. Lowness here means that the several parts of the organisation have been but little specialised for particular functions; and as long as the same part has to perform diversified work, we can perhaps see why it should remain variable, that is, why natural selection should not have preserved or rejected each little deviation of form so carefully as when the part has to serve for some one special purpose. In the same way that a knife which has to cut all sorts of things may be of almost any shape, whilst a tool for some particular purpose must be of some particular shape.

A Part developed in any Species in an extraordinary degree or manner, in comparison with the same Part in allied Species, tends to be highly variable

When we see any organ developed in a remarkable degree in a species, the fair presumption is that it is of high importance to that species; nevertheless it is in this case eminently liable to variation. Why should this be so? On the view that each species has been independently created, with all its parts as we now see them, I can see no explanation. But on the view that groups of species are descended from some other species, and have been modified through natural selection, I think we can obtain some light.

When a part has developed in an extraordinary manner in one species, compared with other species of the same genus, we may conclude that this part has undergone an extraordinary amount of modification since the period when the several species branched off from their common progenitor. This period will seldom be remote, as species rarely endure for more than one geological period. An extraordinary amount of modification implies an unusually large amount of variability, which has accumulated for the benefit of the species. But as the variability of the organ has been so great within a period not excessively remote, we can still find more variability in it than in other parts of the organisation which have remained for a much longer period nearly constant.

Specific Characters more Variable than Generic Characters

The principle discussed under the last heading may be applied to our present subject. It is notorious that specific characters are more variable than generic. On the view of each species having been independently created, why should that part of the structure, which differs from the same part in other independently created species of the same genus, be more variable than those parts which are closely alike in the several species? I do not see that any explanation can be given. But on the view that species are only strongly marked and fixed varieties, we might expect to find them still varying in those parts which have varied within a moderately recent period, and which have thus come to differ. The points in which all the species of a genus resemble each other are called generic characters; and these have been inherited from before the period when the several species first branched off from their common progenitor, and subsequently have not varied, or only in a slight degree; hence it is not probable that they should vary at the present day. On the other hand as specific characters have varied and come to differ since the period when the species branched off from a common progenitor, it is probable that they should still often be in some degree variable.

Secondary Sexual Characters Variable

Species of the same group differ from each other more widely in their secondary sexual characters than in other parts of their organisation: compare, for instance, the amount of difference between the males of gallinaceous birds, in which secondary

sexual characters are strongly displayed, with the amount of difference between the females. We can see why they should not have been rendered as constant and uniform as others, for they are accumulated by sexual selection, which is less rigid in its action than ordinary selection as it does not entail death but only gives fewer offspring to the less favoured males.

It is a remarkable fact that the secondary differences between the two sexes of the same species are generally displayed in the very same parts of the organisation in which the species of the same genus differ from each other. Sir J. Lubbock has recently remarked: 'In Pontella, for instance, the sexual characters are afforded mainly by the anterior antennae and by the fifth pair of legs: the specific differences also are principally given by these organs.' This relation has a clear meaning on my view: whatever part of the structure of the common progenitor became variable, variations of this part would, it is highly probable, be taken advantage of by natural and sexual selection in order to fit the several species to their places in the economy of nature, and likewise to fit the two sexes of the same species to each other.

Distinct Species present analogous Variations, so that a Variety of one Species often assumes a Character proper to an allied Species, or reverts to some of the Characters of an early Progenitor

The most distinct breeds of pigeon present sub-varieties with reversed feathers on the head and with feathers on the feet – characters not possessed by the aboriginal rock-pigeon; these then are analogous variations in distinct races. The frequent presence of fourteen or sixteen tail-feathers in the pouter is a variation representing the normal structure of another race, the fantail. No one will doubt that all such analogous variations are due to inheritance from a common parent of the same tendency to variation. Many similar cases of analogous variation have been observed in the gourd-family, and in our cereals.

With pigeons, however, we have the occasional appearance in all the breeds of coloured marks characteristic of the parent rock-pigeon. I presume that no one will doubt that this is a case of reversion, and not of a new yet analogous variation appearing in the several breeds. These coloured marks are eminently liable to appear in the crossed offspring of two distinct and differently coloured breeds; and in this case there is nothing in the external conditions of life to cause the reappearance of the marks, beyond the influence of the mere act of crossing on the laws of inheritance.

No doubt it is a very surprising fact that characters should reappear after having been lost for many generations. But when a breed has been crossed only once by some other breed, the off-spring occasionally show for many generations a tendency to revert in character to the foreign breed. After twelve generations the proportion of blood, to use a common expression, from one ancestor is only 1 in 2048; and yet a tendency to reversion is retained by this remnant of foreign blood. In a breed which has not been crossed, but which has lost some character which the progenitor possessed, the tendency to reproduce the lost character might be transmitted for almost any number of generations. When a character which has been lost in a breed reappears after a great number of generations, the most probable hypothesis is that in each successive generation the character in question has been lying latent, and at last, under unknown favourable conditions, is developed. The abstract improbability of such a tendency being transmitted through a vast number of generations is not greater than that of quite useless organs being similarly transmitted.

[*We can now understand in genetic terms how a character might be transmitted through many generations without being expressed, and then suddenly recur in a particular individual. In the simplest case the character might depend on a rare recessive gene, and would only be expressed when an individual receives*

such a recessive gene from both its parents, (see diagram on p. 24). In other cases a whole set of genes acting together, known as a 'gene complex', is needed to produce a character, and these must all be brought together in the same individual for the character to be expressed.]

I will give another curious case, almost certainly one of reversion. The ass sometimes has very distinct transverse bars on its legs, like those on the legs of the zebra; these are plainest in the foal. The stripe on the shoulder is sometimes double, and is very variable in length and outline. Mr. Blyth has seen a specimen of the hemionus [Asiatic

A horse showing stripes on its back. The occasional appearance of stripes on the legs or back in horses, asses and mules suggested to Darwin that the common ancestor of the horse, ass, zebra and quagga must have been striped.

wild ass] with a distinct shoulder stripe, though properly it has none, and I have been informed that the foals of this species are generally striped on the legs, and faintly on the shoulder. The quagga, though barred like a zebra over the body, is without bars on the legs; but Dr. Gray has figured one specimen with very distinct zebra-like bars on the hocks.

With respect to the horse, I have collected cases in England of spinal stripe in horses of the most distinct breeds and of *all* colours: transverse bars on the legs are not rare in duns and mouse-duns; a faint shoulder-stripe may sometimes be seen in duns. In north-west India the Kattywar breed of horses is so generally striped that a horse without stripes is not considered purely bred. The stripes are often plainest in the foal and sometimes disappear in old horses. I have collected cases of leg- and shoulder-stripes in horses of very different breeds from Britain to China, and from Norway in the north to the Malay Archipelago in the south. In all parts of the world these stripes occur far oftenest in duns and mouse-duns.

Let us turn to the effects of crossing species of the horse-genus. The common mule from the ass and horse is particularly apt to have bars on its legs. I once saw a mule with its legs so striped that anyone might have thought it a hybrid-zebra. A hybrid from the ass and hemionus, though the ass only occasionally has stripes on his legs and the hemionus has none and has not even a shoulder-stripe, nevertheless had all four legs barred, and had three short shoulder-stripes and even some zebra-like stripes on the sides of its face.

What are we to say to these several facts? We see distinct species of the horse-genus becoming, by simple variation, striped on the legs like a zebra or striped on the shoulders like an ass. In the horse this tendency is strong whenever a dun tint appears – a tint which approaches the general colouring of the other species. The stripes are not accompanied by any change of form or other new character. This tendency is most strongly displayed in hybrids. Now observe the case of the breeds of pigeons: they are descended from a pigeon of bluish colour, with certain bars and other marks; and when any breed assumes by simple variation a bluish tint, these marks invariably reappear without any other change of form. When the oldest and truest breeds are crossed, we see a strong tendency for the blue tint and marks to reappear in the mongrels. Call the breeds species, and how exactly parallel is the case with that of the species of the horse-genus! For myself, I venture confidently to look back thousands on thousands of generations, and I see an animal striped like a zebra, but perhaps very differently constructed, the common parent of our domestic horse, the ass, the hemionus, quagga, and zebra.

He who believes that each equine species was independently created will, I presume, assert that each species has been created with a tendency to vary in this particular manner, so as often to become striped like the other species of the genus; and that each has been created with a strong tendency, when crossed with species inhabiting distant quarters of the world, to produce hybrids resembling in their stripes, not their own parents, but other species of the genus. To admit this view is, as it seems to me, to reject a real for an unreal, or at least an unknown, cause. It makes the works of God a mere mockery and deception; I would almost as soon believe with the old and ignorant cosmogonists, that fossil shells had never lived, but had been created in stone so as to mock the shells living on the sea-shore.

Chapter 6
Difficulties of the Theory

Long before the reader has arrived at this part of my work, a crowd of difficulties will have occurred to him. Some of them are so serious that to this day I can hardly reflect on them without being in some degree staggered; but, to the best of my judgment, the greater number are only apparent, and those that are real are not, I think, fatal to the theory.

These difficulties may be classed under the following heads: first, why, if species have descended from other species by fine gradations, do we not everywhere see innumerable transitional forms? Why is not all nature in confusion, instead of the species being well defined? Secondly, is it possible that an animal having, for instance, the structure and habits of a bat, could have been formed by the modification of some other widely different animal? Thirdly, can instincts be acquired and modified through natural selection? What shall we say to the instinct which leads the bee to make cells, and which has practically anticipated the discoveries of profound mathematicians? Fourthly, how can we account for species, when crossed, being sterile and producing sterile offspring, whereas, when varieties are crossed, their fertility is unimpaired?

The two first heads will here be discussed; some miscellaneous objections in the following chapter; Instinct and Hybridism in the two succeeding chapters.

On the Absence or Rarity of Transitional Varieties

As natural selection acts solely by the preservation of profitable modifications, each new form will tend in a fully stocked country to take the place of its own less improved parent-form and other less-favoured forms with which it comes into competition. Hence both the parent and all transitional varieties will generally have been exterminated by the formation and perfection of the new form.

But why do we not find innumerable transitional forms embedded in countless numbers in the crust of the earth? The answer mainly lies in the record being incomparably less perfect than is generally supposed. The crust of the earth is a vast museum; but the natural collections have been imperfectly made, and only at long intervals of time.

In travelling from north to south over a continent, we generally meet at successive intervals closely allied species evidently filling nearly the same place in the natural economy. These representative species often meet and interlock; and as the one becomes rarer, the other becomes more frequent, till the one replaces the other. If we compare these species where they intermingle, they are generally as distinct from each other as are specimens from the metropolis inhabited by each. These allied species are descended from a common parent; and during the process of modification each has become adapted to its own region, and has supplanted its parent-form and all transitional varieties. Hence we ought not to expect to meet with numerous transitional varieties in each region. But in the intermediate region, having intermediate conditions of life, why do we not now find linking intermediate varieties?

In the first place we should be extremely cautious in inferring, because an area is now continuous, that it has been continuous during a long period. Geology would lead us to believe that most continents have been broken up into islands even during the later Tertiary periods; and in such islands distinct species might have been separately formed without the possibility of intermediate varieties existing in the intermediate zones. But I will pass over this way of escaping from the difficulty; for I believe that many perfectly defined species have been formed on strictly continuous areas.

We generally find species tolerably numerous over a large territory, then becoming abruptly rarer on the confines, and finally disappearing. Hence the neutral territory between two representative species is generally narrow in comparison with the territory proper to each. As varieties do not essentially differ from species, the same rule will probably apply to both: if we take a varying species inhabiting a very large area, we shall have to adapt two varieties to two large areas, and a third variety to a narrow intermediate zone. The intermediate variety, consequently, will exist in lesser numbers from inhabiting a lesser area.

But any form existing in lesser numbers would, during fluctuations in the number of its enemies or prey, or in the nature of the seasons, run a greater chance of being exterminated than one existing in large numbers; and the intermediate form would be liable to the inroads of closely allied forms existing on both sides. But it is far more important that the forms existing in larger numbers will have a better chance of presenting favourable variations for natural selection to seize on, than will the rarer forms. Hence the more common forms will tend to supplant the less common forms, for these will be the most slowly modified and improved.

On the Origin and Transitions of Organic Beings with peculiar Habits and Structure

It has been asked by opponents how a land carnivorous animal could have been converted into one with aquatic habits; for how could the animal in its transitional state have subsisted? It would be easy to show that there now exist carnivorous animals presenting close intermediate grades from strictly terrestrial to aquatic habits. Look at the *Mustela vison* [sea otter] of North America, which has webbed feet and resembles an otter in its fur, short legs, and form of tail. During the summer this animal dives for and preys on fish, but during the winter it leaves the frozen waters and preys, like other polecats, on mice and land animals. If a different case had been taken, and it had been asked how an insectivorous quadruped could possibly have been converted into a flying bat, the question would have been far more difficult to answer. Yet I think such difficulties have little weight.

Look at the family of squirrels; here we have the finest gradation from animals with their tails only slightly flattened or with the posterior part of their bodies rather wide and the skin on their flanks rather full, to the flying squirrels, which have their limbs and the base of the tail united by a broad expanse of skin which serves as a parachute and allows them to glide to an astonishing distance from tree to tree. Each structure is of use to each kind of squirrel in its own country, by enabling it to escape birds or beasts of prey, to collect food more quickly, or to lessen the danger from occasional falls. But it does not follow that the structure of each squirrel is the best under all possible conditions. Let the climate and vegetation change, let competing rodents or new beasts of prey immigrate, or old ones become modified, and some of the squirrels would decrease in numbers or become exterminated unless they improved in structure. Therefore I see no difficulty in the preservation of individuals with fuller and fuller flank-membranes, each modification being useful,

Olympic flying squirrels, painted by J. J. Audubon. They do not fly in the strict sense, but glide by means of the flaps of skin between their limbs.

until a perfect flying squirrel be produced.

Now look at the so-called flying lemur, which formerly was ranked amongst bats but is now believed to belong to the Insectivora. An extremely wide flank-membrane stretches from the corners of the jaw to the tail, and includes the limbs with the elongated fingers. Although no graduated links of structure, fitted for gliding through the air, now connect the flying lemur with the other Insectivora, yet there is no difficulty in supposing that such links formerly existed. Nor can I see any difficulty in believing that the membrane-connected fingers and fore-arm of the flying lemur might have been greatly lengthened by natural selection; and this, as far as the organs of flight are concerned, would have converted the animal into a bat. In certain bats in which the wing-membrane extends from the top of the shoulder to the tail and includes the hind-legs, we perhaps see traces of an apparatus originally fitted for gliding through the air rather than for flight.

If about a dozen genera of birds were to become extinct, who would surmise that birds might have existed which used their wings solely as flappers, like the logger-headed duck [steamer duck], as fins in the water and as front-legs on the land, like the penguin, and as sails, like the ostrich? Yet the structure of each bird is good for the conditions of life to which it is exposed. But it is not necessarily the best possible under all possible conditions. These grades of wing-structure show what diversified means of transition are at least possible.

Seeing that members of such water-breathing classes as the Crustacea and Mollusca are adapted to live on land, and that we have flying birds and mammals, flying insects, and formerly had flying reptiles, it is conceivable that flying fish, which now glide far through the air, slightly rising and turning by the aid of their fluttering fins, might have been modified into perfectly winged animals. If this had been effected, who would have ever imagined that in an early transitional state they had been the inhabitants of the open ocean, and

A flying lemur with young. This unique animal glides in the manner of flying squirrels, but has a far more extensive gliding membrane. Darwin suggested that the flying lemur might represent a link between the Insectivores and the bats, but it is doubtful that this is the case.

had used their incipient organs of flight exclusively, as far as we know, to escape being devoured by other fish?

When we see any structure highly perfected, as the wings of a bird, we should bear in mind that animals displaying early transitional grades of the structure will seldom have survived to the present day, for they will have been supplanted by their successors, rendered more perfect through natural selection. Furthermore, transitional states will rarely have developed in great numbers and many subordinate forms. To return to our imaginary illustration, fishes capable of true flight would not have developed many subordinate forms for taking prey of many kinds in many ways, on land and in water, until their organs of flight had come to a high stage of perfection, so as to have given them a decided advantage over other animals in the battle for life. Hence the chance of discovering transitional species in a fossil condition will always be less.

I will now give instances of diversified and changed habits in individuals of the same species. Many British insects now feed on exotic plants or artificial substances. I have often watched a tyrant flycatcher in South America hovering over one spot and then proceeding to another, like a kestrel, and at other times standing stationary on the margin of water, and then dashing into it like a kingfisher at a fish. In our own country the larger titmouse [great tit] may be seen climbing branches, almost like a creeper; it sometimes, like a shrike, kills small birds by blows on the head; and I have many times seen it hammering the seeds of the yew on a branch, and thus breaking them like a nuthatch. In North America the black bear has been seen swimming for hours with widely open mouth, thus catching, almost like a whale, insects in the water.

As we sometimes see individuals following habits different from those proper to their species, we might expect that such individuals would occasionally give rise to new species, having anomalous habits and with their structure modified from that of their type. And such instances occur in nature. Can a more striking instance of adaptation be given than that of a woodpecker for climbing trees and seizing insects in the chinks of the bark? Yet in North America there are woodpeckers which feed largely on fruit, and others with elongated wings which chase insects on the wing. As another illustration of the varied habits of this genus, a Mexican woodpecker has been described as boring holes into hardwood in order to lay up a store of acorns.

Petrels are the most aerial and oceanic birds, but

The acorn woodpecker from Mexico which uses its wood-boring abilities to make holes in which to store acorns. The trunk of the woodpecker's storage tree is shown above.

The dipper, known in Darwin's time as a water-ouzel, feeds on water insects, swimming underwater by means of its wings. It shows certain adaptations to its way of life, for example well waterproofed feathers, but none of the features commonly associated with water birds, such as webbed feet. *'... the acutest observer, by examining its dead body would never have suspected its sub-aquatic habits.'*

He who believes that each being has been created must occasionally have felt surprise when he has met with an animal having habits and structure not in agreement. What can be plainer than that the webbed feet of ducks and geese are formed for swimming? Yet there are upland geese with webbed feet which rarely go near the water; and no one except Audubon has seen the frigate-bird, which has all its four toes webbed, alight on the surface of the ocean. What seems plainer than that the long toes, not furnished with membrane, of the Grallatores are formed for walking over swamps and floating plants? The water-hen [moorhen] and landrail [corncrake] are members of this order, yet the first is nearly as aquatic as the coot, and the second nearly as terrestrial as the quail or partridge. In such cases habits have changed without a corresponding change of structure. The webbed feet of the upland goose may be said to have become almost rudimentary in function, though not in structure. In the frigate-bird, the deeply scooped membrane between the toes shows that structure has begun to change.

He who believes in the struggle for existence will acknowledge that every organic being is constantly endeavouring to increase in numbers; and that if any being varies ever so little, either in habits or structure, and thus gains an advantage over some other inhabitant, it will seize on the place of that inhabitant, however different that may be from its own place. Hence it will cause him no surprise that there should be geese and frigate-birds with webbed feet, living on the dry land and rarely alighting on the water; that there should be

in the quiet sounds of Tierra del Fuego there is a species which, in its general habits, in its astonishing power of diving, in its manner of swimming and flying would be mistaken by anyone for an auk or a grebe; nevertheless it is essentially a petrel, but with many parts of its organisation profoundly modified in relation to its new habits of life. In the case of the water-ouzel [dipper] the acutest observer by examining its dead body would never have suspected its sub-aquatic habits; yet this bird, which is allied to the thrush family, subsists by diving – using its wings under water, and grasping stones with its feet. All Hymenopterous insects are terrestrial, excepting the genus *Proctotrupes*, which is aquatic in its habits; it often enters the water and dives about by the use not of its legs but of its wings, and remains as long as four hours beneath the surface; yet it exhibits no modification in structure in accordance with its abnormal habits.

'What seems plainer than that the long toes, not furnished with membrane, of the Grallatores are formed for walking over swamps and floating plants? – the water-hen and landrail are members of this order, yet the first is nearly as aquatic as the coot, and the second nearly as terrestrial as the quail or partridge. In such cases habits have changed without a corresponding change of structure.' These paintings by John Gould show a moorhen (known in Darwin's time as a water-hen) above, and a corncrake (Darwin's landrail) below. Most species of the Grallatores inhabit swamps or pond margins, walking on the leaves of water plants as the moorhen chicks are doing in the upper picture.

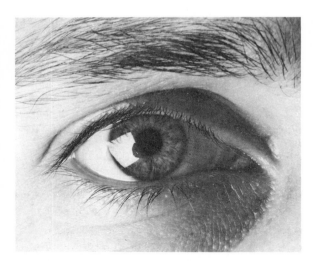

Man

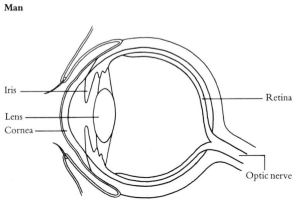

Insect

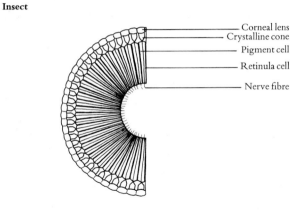

Octopus

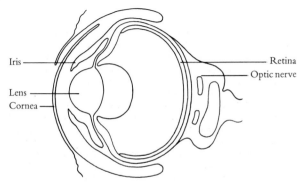

long-toed corncrakes living in meadows instead of in swamps, that there should be diving thrushes and diving Hymenoptera, and petrels with the habits of auks.

Organs of extreme Perfection and Complication

To suppose that the eye with all its inimitable contrivances for adjusting the focus to different distances, for admitting different amounts of light, and for the correction of spherical and chromatic aberration could have been formed by natural selection seems, I freely confess, absurd in the highest degree. But when it was first said that the sun stood still and the world turned round, the common sense of mankind declared the doctrine false.

If numerous gradations from a simple and imperfect eye to one complex and perfect can be shown to exist, each grade being useful to its possessor, and if the eye varies and the variations be inherited, then the difficulty of believing that a perfect and complex eye could be formed by natural selection should not be considered as subversive of the theory. How a nerve comes to be sensitive to light hardly concerns us more than how life itself originated; but as some of the lowest organisms, in which nerves cannot be detected, are capable of perceiving light, it does not seem impossible that certain elements in them

should become aggregated and developed into nerves endowed with this special sensibility.

In searching for the gradations through which an organ in any species has been perfected, we ought to look exclusively to its progenitors; but this is scarcely ever possible, and we are forced to look to other species and genera of the same group in order to see what gradations are possible. Even the state of the same organ in distinct classes may incidentally throw light on the steps by which it has been perfected.

The simplest organ which can be called an eye consists of an optic nerve, surrounded by pigment-cells and covered by translucent skin, but without any lens. We may, however, descend even lower and find aggregates of pigment-cells without any nerves. Eyes of the above nature are not capable of distinct vision, and serve only to distinguish light from darkness. In certain star-fishes, small depressions in the layer of pigment which surrounds the nerve are filled with transparent gelatinous matter, projecting with a convex surface like the cornea in the higher animals. This serves not to form an image but only to concentrate the luminous rays and render their perception more easy. In this concentration we gain the most important step towards the formation of a true, picture-forming eye, for we have only to place the naked extremity of the optic nerve at the right distance from the concentrating apparatus, and an image will be formed on it.

In the great class of Articulata [Arthropoda], we start from an optic nerve simply coated with pigment, the latter sometimes forming a sort of pupil, but destitute of a lens. With insects the numerous facets on the cornea of their compound eyes form true lenses, and the cones include modified nervous filaments. These organs in the Articulata are much diversified.

When we reflect on the wide, diversified and graduated range of structure in the eyes of the lower animals, and bear in mind how small the number of all living forms must be in comparison

A complex eye has evolved in several groups of animals. Vertebrates such as man have an eye that consists of a single lens adjustable by muscle action to focus an image on the retina from which optic nerves carry messages to the brain; the size of the iris can be adjusted to let in more or less light. Cephalopod molluscs, that is the octopuses, squids and cuttlefish, have quite independently evolved an eye which is remarkably similar to that of vertebrates: an example of convergent evolution. The compound eyes of insects, however, are very different. These eyes are made up of numerous units, each of which has a corneal lens, and a crystalline cone which also acts as a lens. The image is received by the retinula cells and passed along a nerve fibre to the brain. In some insects the pigment cells between the units of the compound eye can move nearer to the surface to reduce the amount of light entering the eye.

with those which have become extinct, the difficulty ceases to be very great in believing that natural selection may have converted the simple apparatus of an optic nerve, coated with pigment and invested by transparent membrane, into an optical instrument as perfect as is possessed by any member of the Articulate Class.

It has been objected that in order to modify the eye and still preserve it as a perfect instrument, many changes would have to be effected simultaneously. But as I have shown on the variation of domestic animals, it is not necessary to suppose that the modifications were all simultaneous, if they were extremely slight and gradual. Different kinds of modification would, also, serve for the same general purpose: as Mr. Wallace has remarked, 'if a lens has too short or too long a focus, it may be amended either by an alteration of curvature or an alteration of density; if the curvature be irregular, and the rays do not converge to a point, then any increased regularity of curvature will be an improvement. So the contraction of the iris and the muscular movements of the eye are neither of them essential to vision, but only improvements which might have been added and perfected at any stage of the construction of the instrument.'

Within the highest division of the animal kingdom, namely the Vertebrata, we can start from an eye so simple that it consists, as in the lancelet, of a little sack of transparent skin, furnished with a nerve and lined with pigment, but destitute of any other apparatus. In fishes and reptiles the range of gradations is very great. It is a significant fact that even in man the beautiful crystalline lens is formed in the embryo by an accumulation of epidermic cells, lying in a sacklike fold of the skin; and the vitreous body is formed from embryonic subcutaneous tissue.

Modes of Transition

If it could be demonstrated that any complex organ existed which could not possibly have been formed by numerous, successive, slight modifications, my theory would absolutely break down. But I can find out no such case.

We should be extremely cautious in concluding that an organ could not have been formed by transitional gradations of some kind. Amongst the lower animals the same organ can perform at the same time wholly distinct functions; thus in the larva of the dragon-fly and in the fish Cobites [weatherfish or pond loach] the alimentary canal respires, digests, and excretes. In the Hydra, the animal may be turned inside out, and the exterior surface will then digest and the stomach respire. In such cases natural selection might specialise the organ which had previously performed two functions for one function alone, and thus by insensible steps change its nature.

Again, two distinct organs may simultaneously perform in the same individual the same function, and this is an extremely important means of transition: to give one instance, there are fish with gills that breathe air dissolved in water, at the same time that they breathe free air in their swimbladders. Plants climb by three distinct means, by spirally twining, clasping a support with their tendrils, and the emission of aerial rootlets; these are usually found in distinct groups, but some species exhibit two of the means, or even all three, combined in the same individual. In such cases one of the organs might readily be modified to perform all the work, being aided during the modification by the other organ; then this other organ might be modified for some quite distinct purpose.

The illustration of the swimbladder shows us clearly that an organ originally constructed for one purpose, namely flotation, may be converted into one for a different purpose, namely respiration. The swimbladder has also been worked in as an accessory auditory organ of certain fishes. Its similarity in position and structure with the lungs of the higher vertebrate animals leaves no doubt that the swimbladder has actually been converted into lungs. [*It is now believed that lung-like structures*

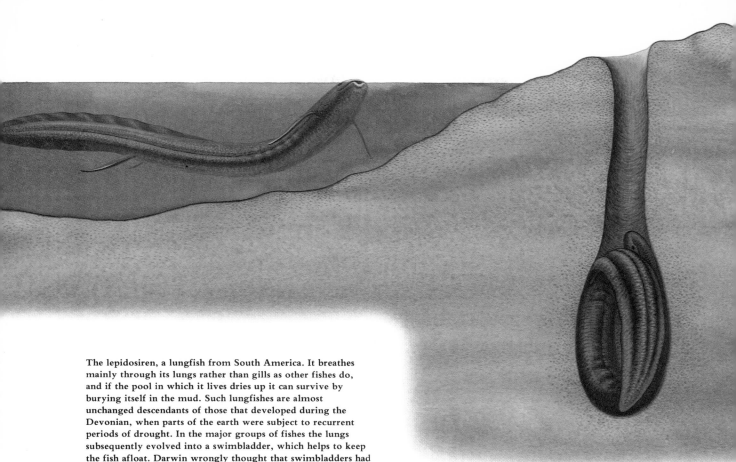

The lepidosiren, a lungfish from South America. It breathes mainly through its lungs rather than gills as other fishes do, and if the pool in which it lives dries up it can survive by burying itself in the mud. Such lungfishes are almost unchanged descendants of those that developed during the Devonian, when parts of the earth were subject to recurrent periods of drought. In the major groups of fishes the lungs subsequently evolved into a swimbladder, which helps to keep the fish afloat. Darwin wrongly thought that swimbladders had evolved first and that lungs had then developed from them.

evolved first in fish, and that swimbladders developed from these lungs.]

According to this view all vertebrate animals with true lungs are descended from an unknown prototype furnished with a swimbladder. We can thus understand the strange fact that every particle of food we swallow has to pass over the orifice of the trachea, with some risk of falling into the lungs, notwithstanding the beautiful contrivance by which the glottis is closed. In the higher Vertebrata the gills have wholly disappeared – but in the embryo the slits on the sides of the neck and the loop-like course of the arteries still mark their former position. But it is conceivable that the now utterly lost gills might have been gradually worked in by natural selection for some distinct purpose: for the wings of insects are developed from the tracheae, which once served for respiration.

There is another possible mode of transition: some animals are capable of reproduction at a very early age, before they have acquired their perfect characters; if this power became thoroughly well developed in a species, it seems probable that the adult stage of development would sooner or later be lost; and in this case the character of the species would be greatly changed and degraded. Again, not a few animals, after arriving at maturity, go on changing during nearly their whole lives. With mammals the form of the skull is often much altered with age. The teeth of certain lizards change much in shape with advancing years; with crustaceans some important parts assume a new character after maturity. In such cases, if the age for reproduction were retarded, the character of the species, at least in its adult state, would be modified; nor is it improbable that the earlier stages of development would in some cases be hurried through and finally lost.

'There is another possible mode of transition: some animals are capable of reproduction at a very early age, before they have acquired their perfect characters; if this power became thoroughly developed in a species, it seems probable that the adult stage of development would sooner or later be lost; and in this case the character of the species would be greatly changed and degraded.' This evolutionary mechanism, often referred to as *neoteny*, is regarded as highly important by modern evolutionary biologists. For example, in the evolution of man from the apes, several juvenile features have been retained into adulthood. The axolotl is an interesting example of an animal that reproduces while still in its juvenile form. It is a salamander that spends most of its life in water and retains the features of the tadpole stage, including external gills, although it grows up to nine inches long. Only if its pond dries up will it metamorphose into a typical adult salamander in order to migrate to another pond.

Special Difficulties of the Theory of Natural Selection

The electric organs of fishes offer a special difficulty; for it is impossible to conceive by what steps these wondrous organs have been produced. We do not even know of what use they are. In the Gymnotus and Torpedo they serve as powerful means of defence, and perhaps for securing prey; yet in the Ray an analogous organ in the tail manifests so little electricity, even when the animal is greatly irritated, that it can hardly be of any use for these purposes. It is generally admitted that there exists between these organs and ordinary muscle a close analogy, and that muscular contraction is accompanied by an electrical discharge. Beyond this we cannot at present go in the way of explanation; but it would be extremely bold to maintain that no transitions are possible by which these organs might have been gradually developed.

These organs appear at first to offer another and far more serious difficulty; for they occur in about a dozen kinds of fish, of which several are widely remote in their affinities. When the same organ is found in several members of the same class, we may generally attribute its presence to inheritance from a common ancestor. But we see in the several fishes provided with electric organs that these are situated in different parts of the body, that they differ in construction, as in the arrangement of the plates, and in the process by which the electricity is excited, and lastly in being supplied with nerves proceeding from different sources. Consequently there is no reason to suppose that they have been inherited from a common progenitor; for had this been the case they would have closely resembled each other in all respects. Thus the difficulty of an organ, apparently the same, arising in several remotely allied species disappears, leaving only the lesser difficulty, namely by what steps these organs have been developed in each group of fishes.

The luminous organs which occur in a few insects belonging to widely different families offer an almost exact parallel. In plants, the very curious contrivance of a mass of pollen-grains borne on a foot-stalk with an adhesive gland is apparently the same in *Orchis* and *Asclepias* – genera almost as remote as is possible amongst flowering plants. In all cases of beings far removed from each other in the scale of organisation which are furnished with similar and peculiar organs, it will be found that fundamental differences between the organs can always be detected. For instance, the eyes of cephalopods and of vertebrate animals appear wonderfully alike. But an organ for vision must be formed of transparent tissue and include some sort of lens for throwing an image at the back of a darkened chamber. Beyond this superficial resemblance, there is hardly any real similarity between the eyes of cephalopods and vertebrates. The crystalline lens in the higher cuttle-fish consists of two parts, placed one behind the other like two lenses, both having a very different structure and disposition to what occurs in the Vertebrata. The retina is wholly different, with an actual inversion of the elemental parts and with a large nervous ganglion included within the membranes of the eye. The relations of the muscles are as different as possible, and so in other points. As two men have sometimes independently hit on the same invention, so natural selection, working for the good of each being, has produced similar organs, as far as function is concerned, in distinct organic beings.

On the other hand, it is a common rule throughout nature that the same end should be gained, even sometimes in the case of closely related beings, by the most diversified means. How differently constructed is the feathered wing of a bird and the membrane-covered wing of a bat. Seeds are disseminated by their capsule being converted into a light balloon-like envelope, by being embedded in pulp or flesh and rendered nutritious, as well as conspicuously coloured, so as to attract and be devoured by birds, by having hooks so as to adhere to the fur of quadrupeds, and by being furnished with wings and plumes so as to be wafted by every breeze. Some authors maintain that organic beings have been formed in many ways for the sake of mere variety, almost like toys in a shop, but such a view of nature is incredible.

With plants some aid is necessary for their fertilisation. With several kinds this is effected by the pollen-grains being blown by the wind through mere chance on to the stigma. An almost equally simple plan occurs in many plants in which a symmetrical flower secretes a few drops of nectar, and is consequently visited by insects; and these carry the pollen from the anthers to the stigma.

From this simple stage we may pass through an inexhaustible number of contrivances, all for the same purpose and effected in essentially the same manner, but entailing changes in every part of the flower. The nectar may be stored in variously shaped receptacles, with the stamens and pistils modified in many ways, sometimes forming trap-like contrivances and sometimes capable of neatly adapted movements through irritability or elasticity. From such structures we may advance to the extraordinary adaptation of the *Coryanthes*. This orchid has part of its labellum or lower lip hollowed out into a great bucket, into which drops of almost pure water continually fall from two secreting horns; and when the bucket is half full the water overflows by a spout on one side. The basal part of the labellum stands over the bucket, and is itself hollowed out into a sort of chamber with two lateral entrances; within this chamber there are curious fleshy ridges. Crowds of large humble-bees [bumble-bees] visit the gigantic flowers of this orchid, not to suck nectar but to gnaw off the ridges within the chamber; in doing this they frequently push each other into the bucket, and their wings being thus wetted they cannot fly away, but are compelled to crawl out through the passage formed by the overflow. The passage is narrow and is roofed over by the column, so that a bee, in forcing its way out, first rubs its back against the viscid stigma and then against the viscid glands of the pollen-masses. The pollen-masses are thus glued to the back of the bee which first happens to crawl out through the passage of a lately expanded flower. When the bee, thus provided, flies to another flower and is pushed into the bucket and then crawls out by the passage, the pollen-mass comes first into contact with the viscid stigma and adhere to it, and the flower is fertilised. Now at last we see the full use of every part of the flower.

How, in the foregoing instances, can we understand the graduated scale of complexity and the
multifarious means for gaining the same end? The
answer is that when two forms vary, which already
differ from each other, the variability will not be
the same, and the results obtained through natural
selection for the same general purpose will not be
the same. Every highly developed organism has
passed through many changes, and each modification may be again and again altered. Hence the
structure of each part of each species is the sum of
many inherited changes.

Finally then, considering how small the proportion of living and known forms is to the extinct
and unknown, I have been astonished how rarely
an organ can be named towards which no transitional grade is known to lead. It certainly is true
that new organs appearing as if created for some
special purpose rarely or never appear in any
being. Or as Milne Edwards has well expressed it,
Nature is prodigal in variety, but niggard in
innovation. Why, on the theory of Creation,
should there be so much variety and so little real
novelty? Why should all the parts and organs of
many independent beings, each supposed to have
been separately created for its proper place in
nature, be so commonly linked together by
graduated steps? Why should not Nature take a
sudden leap from structure to structure? On the
theory of natural selection, we can clearly understand why she should not; for natural selection acts
only by taking advantage of slight successive
variations; she can never take a great and sudden
leap, but must advance by short and sure, though
slow steps.

Organs of little apparent Importance, as affected by Natural Selection

As natural selection acts by life and death I have
sometimes felt great difficulty in understanding
the origin or formation of parts of little importance.

We are much too ignorant in regard to the whole
economy of any organic being to say what slight

modifications would be of importance or not. I
have given instances of very trifling characters,
such as the down on fruit and the colour of its flesh,
which, from being correlated with constitutional
differences or from determining the attacks of
insects, might be acted on by natural selection. The
tail of the giraffe looks like a fly-flapper; and it
seems at first incredible that this could have been
adapted by successive modifications for so trifling
an object as to drive away flies; yet we know that
the distribution and existence of cattle in South
America absolutely depend on their power of
resisting the attacks of insects. It is not that the
larger quadrupeds are actually destroyed by
flies, but they are incessantly harassed and their
strength reduced, so that they are more subject to
disease, or not so well enabled to search for food
or to escape from beasts of prey

Organs now of trifling importance have probably in some cases been of high importance to a
progenitor, and have been transmitted to existing
species, although now of very slight use. In other
cases, we may easily err in attributing importance
to characters, and in believing that they have been
developed through natural selection. We must by
no means overlook the effects of spontaneous
variations, of reversion to long-lost characters, of
the complex laws of growth, such as of correlation,
compensation, of the pressure of one part on
another, and finally of sexual selection.

If green woodpeckers alone had existed, and
there were no black or pied kinds, I dare say we
should have thought that the green colour was a
beautiful adaptation to conceal this bird from its
enemies; and consequently that it was a character
of importance, and had been acquired through
natural selection; as it is, the colour is probably in
chief part due to sexual selection. The naked skin
on the head of a vulture is generally considered as a
direct adaptation for wallowing in putridity; and
so it may be, but the skin on the head of the clean-
feeding male turkey is likewise naked. The sutures
in the skulls of young mammals have been ad-

vanced as a beautiful adaptation for aiding par-turition, and no doubt they facilitate, or may be indispensable for this act; but as sutures occur in the skulls of young birds and reptiles, which have only to escape from a broken egg, this structure must have arisen from the laws of growth, and then been taken advantage of in the higher animals.

Utilitarian Doctrine, how far true: Beauty, how acquired

Some naturalists have lately protested against the utilitarian doctrine that every detail of structure has been produced for the good of its possessor. They believe that many structures have been created for the sake of beauty, to delight man or the Creator, or for the sake of mere variety, a view already discussed. I fully admit that many struc-tures are of no direct use to their possessors, but the chief part of the organisation of every living creature is due to inheritance; and consequently many structures have now no very close relation to present habits of life. Thus we can hardly believe that the webbed feet of the upland goose or of the frigate-bird are of special use to these birds. We may safely attribute these structures to inheritance, and webbed feet no doubt were useful to their progenitors.

With respect to the belief that organic beings have been created beautiful for the delight of man, I may first remark that the sense of beauty obviously depends on the nature of the mind, irrespective of any real quality in the admired object, and that the idea of what is beautiful is not innate or unalterable. We see this in men of dif-ferent races admiring an entirely different standard of beauty in their women. If beautiful objects had been created solely for man's gratification, it

Beauty, the Creationists insisted, could not have evolved by natural selection. But what is beauty? Darwin argues that it is largely subjective: 'We see this in the men of different races admiring an entirely different standard of beauty in their women.' A Masai girl from Kenya (above) and a girl of the Nuba tribe from the Sudan (below).

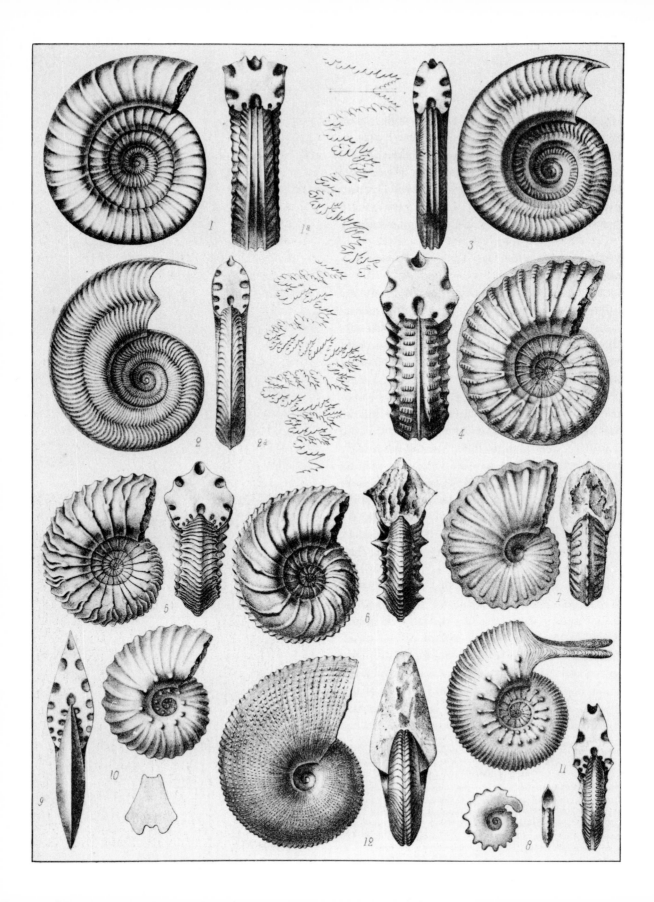

ought to be shown that before man appeared there was less beauty on the earth than since he came. Were the beautiful volute and cone shells of the Eocene epoch and the gracefully sculptured ammonites of the Secondary [Mesozoic] period created that man might ages afterwards admire them in his cabinet? Few objects are more beautiful than the minute cases of the Diatomaceae: were these created that they might be examined and admired under the higher powers of the microscope? The beauty in this case is apparently wholly due to symmetry of growth. Flowers rank amongst the most beautiful productions of nature; but they have been rendered conspicuous in contrast with the green leaves, and in consequence beautiful, so that they may be easily observed by insects. When a flower is fertilised by the wind it never has a gaily coloured corolla. Several plants habitually produce two kinds of flowers: one kind open and coloured so as to attract insects; the other closed, not coloured, destitute of nectar, and never visited by insects. Hence we may conclude that, if insects had not been developed, our plants would not have beautiful flowers, but would produce only such poor flowers as we see on our fir and ash trees, on grasses, spinach and nettles, which are all fertilised through wind. Similarly the beauty of a ripe strawberry or cherry or the scarlet berries of the holly serves merely as a guide to birds and beasts, in order that the fruit may be devoured and the manured seeds disseminated.

On the other hand, I willingly admit that a great number of male animals, as all our most gorgeous birds, some fishes, reptiles and mammals, and a host of magnificently coloured butterflies, have been rendered beautiful for beauty's sake: but this has been effected through sexual selection, that is, by the more beautiful males having been continually preferred by the females, and not for the delight of man. So it is with the music of birds. We may infer from this a nearly similar taste for beautiful colours and musical sounds through a large part of the animal kingdom. When the female is as beautifully coloured as the male, which is not rarely the case with birds and butterflies, the cause apparently lies in the colours acquired through sexual selection having been transmitted to both sexes, instead of to the males alone. How the sense of beauty was first developed in the mind of man and of the lower animals, is a very obscure subject. Habit appears to have come into play; but there must be some fundamental cause in the constitution of the nervous system in each species.

(left) *'If beautiful objects had been created solely for man's gratification, it ought to be shown that before man appeared there was less beauty on the earth than since he came. Were the gracefully sculptured ammonites of the Secondary period created that man might ages afterwards admire them in his cabinet?'* **This engraving shows a selection of ammonites, a large group of marine animals that became extinct at about the same time as the dinosaurs.**

(right) **Many microscopic organisms are extremely beautiful. Darwin asks:** *'. . . were these created that they might be examined and admired under the higher powers of the microscope? The beauty in this case is apparently wholly due to symmetry of growth.'*

'Natural selection cannot produce any modification in a species exclusively for the good of another species ... Some authors suppose that the rattlesnake is furnished with a rattle for its own injury, namely to warn its prey ... It is a much more probable view that the rattlesnake uses its rattle and the cobra expands its frill, and the puff-adder swells whilst hissing so loudly and harshly, in order to alarm the many birds and beasts which attack even the most venomous species.'

Natural selection cannot produce any modification in a species exclusively for the good of another species, but it does often produce structures for the direct injury of other animals, as in the fang of the adder and the ovipositor of the ichneumon, by which its eggs are deposited in the living bodies of other insects. Some authors suppose that the rattlesnake is furnished with a rattle for its own injury, namely to warn its prey. I would as soon believe that the cat curls the end of its tail when preparing to spring in order to warn the doomed mouse. It is a much more probable view that the rattlesnake uses its rattle and the cobra expands its frill, and the puff-adder swells whilst hissing so loudly and harshly, in order to alarm the many birds and beasts which attack even the most venomous species. Snakes act on the same principle which makes the hen ruffle her feathers and expand her wings when a dog approaches her chicks.

Natural selection will never produce in a being any structure more injurious than beneficial to that being. If a balance be struck between the good and evil caused by each part, each will be found on the whole advantageous. Under changing conditions of life, if any part comes to be injurious it will be modified; or if it be not so, the being will become extinct as myriads have become extinct.

Natural selection tends only to make each organic being as perfect as the other inhabitants with which it comes into competition. The endemic productions of New Zealand, for instance, are perfect one compared with another; but they are now rapidly yielding before plants and animals from Europe. We do not always meet with absolute perfection under nature. The correction for the aberration of light is not perfect even in that most perfect organ, the human eye. Can we consider the sting of the bee as perfect, which cannot be withdrawn, and thus inevitably causes the death of the insect by tearing out its viscera?

If the bee's sting existed in a remote progenitor as a boring and serrated instrument, like that in so many members of the same order, and was modified but not perfected for its present purpose, we can perhaps understand how it is that the use of the sting should so often cause the insect's own death: for if on the whole the sting be useful to the community, it will fulfil all the requirements of natural selection, though it may cause the death of some few members. If we admire the truly wonderful power of scent by which the males of many insects find their females, can we admire the production for this single purpose of thousands of drones which are utterly useless to the community for any other purpose, and which are ultimately slaughtered by their industrious and sterile sisters? It may be difficult, but we ought to admire the savage instinctive hatred of the queen-bee, which urges her to destroy the young queens, her daughters, as soon as they are born, or to perish herself in the combat; for undoubtedly this is for the good of the community; and maternal love or maternal hatred, though the latter fortunately is most rare, is all the same to the inexorable principle of natural selection.

Summary: the Law of Unity of Type and of the Conditions of Existence embraced by the Theory of Natural Selection

We have in this chapter discussed some of the difficulties and objections which may be urged against the theory. Many of them are serious; but I think that in the discussion light has been thrown on several facts which on the belief of independent acts of creation are utterly obscure.

It is generally acknowledged that all organic beings have been formed on two great laws – Unity of Type, and the Conditions of Existence. By unity of type is meant that fundamental agreement in structure which we see in organic beings of the same class, and which is quite independent of their habits of life. On my theory, unity of type is explained by unity of descent. The expression of conditions of existence is fully embraced by the principle of natural selection. For natural selection acts by either now adapting the varying parts of each being to its organic and inorganic conditions of life, or by having adapted them during past periods of time.

Chapter 7
Miscellaneous Objections to the Theory of Natural Selection

I will devote this chapter to various miscellaneous objections which have been advanced against my views; but it would be useless to discuss all of them, as many have been made by writers who have not taken the trouble to understand the subject.

It has been argued that, as none of the animals and plants of Egypt have changed during the last three or four thousand years, so probably have none in any part of the world. But the many animals which have remained unchanged since the commencement of the glacial period would have been an incomparably stronger case, for these have been exposed to great changes of climate and have migrated over great distances, whereas in Egypt, during the last several thousand years, the conditions of life have remained uniform. The fact of little modification since the glacial period would have been of some avail against an innate and necessary law of development, but is powerless against the doctrine of natural selection, which implies that when variations of a beneficial nature happen to arise, these will be preserved.

Characters of small functional importance the most constant

A serious objection has been urged by Bronn, namely that many characters appear to be of no service whatever to their possessors, and therefore cannot have been influenced through natural selection; for example the length of the ears and tails in the different species of hares and mice. With respect to plants, Nägeli admits that natural selection has effected much, but insists that the families of plants differ chiefly in morphological characters quite unimportant for the welfare of the species. He consequently believes in an innate tendency towards progressive and more perfect development. He specifies the numerical divisions in the flower parts, the position of the ovules, the shape of the seed when not of any use for dissemination, etc., as cases in which natural selection could not have acted.

There is much force in the above objection. Nevertheless we ought, in the first place, to be extremely cautious in deciding what structures are or have been of use to each species. I may mention that the external ears of the common mouse are supplied in an extraordinary manner with nerves, so that they no doubt serve as tactile organs; hence the length of the ears can hardly be unimportant. With respect to plants, the flowers of orchids present a multitude of curious structures which a few years ago would have been considered without any special function; but they are now known to be of the highest importance for fertilisation by insects. In the second place, it should always be borne in mind that when one part is modified, so will other parts through certain dimly seen causes such as correlation.

Supposed incompetence of natural selection to account for the incipient stages of useful structures

A distinguished zoologist, Mr. St. George Mivart, has recently collected all the objections which have ever been advanced against the theory of selection, as propounded by Mr. Wallace and myself, and has illustrated them with admirable art and force. When thus marshalled, they make a formidable array; and as it forms no part of Mr. Mivart's plan to give the various facts and considerations opposed to his conclusions, no slight effort of reason and memory is left to the reader who may wish to weigh the evidence on both sides.

All Mr. Mivart's objections will be, or have been, considered in the present volume. The one new point which appears to have struck many readers is 'that natural selection is incompetent to account for the incipient stages of useful structures.' This subject is intimately connected with that of the gradation of characters, often accompanied by a change of function: points which were discussed in the last chapter. Nevertheless, I will here consider the most illustrative cases advanced by Mr. Mivart.

The giraffe, by its lofty stature, much-elongated neck, fore-legs, head and tongue has its whole frame beautifully adapted for browsing on the higher branches of trees. It can thus obtain food beyond the reach of the other Ungulata or hoofed animals; and this must be a great advantage during dearths. With the nascent giraffe, the individuals which were the highest browsers and were able during dearths to reach even an inch above the others will often have been preserved. These will have left offspring inheriting the same bodily peculiarities, whilst the individuals less favoured in the same respects will have been the most liable to perish. By this process long continued an ordinary hoofed quadruped might be converted into a giraffe.

To this conclusion Mr. Mivart brings forward two objections. One is that the increased size of the body would obviously require an increased supply of food, and he considers it 'very problematical whether the disadvantages thence arising would not, in times of scarcity, more than counterbalance the advantages.' But as the giraffe does actually exist in large numbers in South Africa, and as some of the largest antelopes in the world, taller than an ox, abound there, why should we doubt that intermediate gradations could formerly have existed there? Assuredly the being able to reach food untouched by other quadrupeds would have been of some advantage to the nascent giraffe.

Mr. Mivart then asks, if natural selection be so potent, and if high browsing be so great an advantage, why has not any other quadruped acquired a long neck and lofty stature? The answer is not difficult: in every meadow in England we see the lower branches of trees trimmed by browsing horses or cattle; what advantage would it be to sheep to acquire slightly longer necks? In every district one kind of animal will be able to browse higher than the others; this one kind alone could have its neck elongated. In South Africa the competition for browsing on the higher branches must be between giraffe and giraffe, and not with other animals.

Why, in other quarters of the world, various animals have not acquired an elongated neck cannot be distinctly answered; but we can see, in a general manner, various causes. To reach the foliage at a considerable height (without climbing) implies great bulk; and we know that some areas support singularly few large quadrupeds, for instance South America, though South Africa abounds with them. Why this should be so, we do not know; nor why the later Tertiary periods should have been much more favourable for their existence than the present time.

In order that an animal should acquire some specially developed structure, it is almost indispensable that several other parts should be modified. Although every part of the body varies, the necessary parts do not always vary in the right direction and to the right degree; also, natural selection is a slow process, and the favourable conditions must long endure in order that any marked effect should thus be produced. A transition of structure, with each step beneficial, is a highly complex affair; there is nothing strange in a transition not having occurred in any particular case.

More than one writer has asked, why have some

which they feed offer an excellent instance. On this head, Mr. Mivart remarks, 'As, according to Mr. Darwin's theory, there is a constant tendency to indefinite variation, it is difficult to see how such indefinite oscillations can ever build up a sufficiently appreciable resemblance to a leaf or other object for Natural Selection to seize upon and perpetuate.'

But in the foregoing cases the insects in their original state no doubt presented some rude and accidental resemblance to an object commonly

(opposite) 'The giraffe, by its lofty stature, much-elongated neck, forelegs, head and tongue has its whole frame beautifully adapted for browsing on the higher branches of trees. It can thus obtain food beyond the reach of the other Ungulata or hoofed animals; and this must be a great advantage during dearths. With the nascent giraffe, the individuals which were the highest browsers and were able during dearths to reach even an inch above the others will often have been preserved. These will have left offspring inheriting the same bodily peculiarities, whilst the individuals less favoured in the same respects will have been the most liable to perish. By this process long continued an ordinary hoofed quadruped might be converted into a giraffe.'

(above) '... if natural selection be so potent, and if high browsing be so great an advantage, why has not any other quadruped acquired a long neck and lofty stature? The answer is not difficult: in every meadow in England we see the lower branches of trees trimmed by browsing horses or cattle; what advantage would it be to sheep to acquire slightly longer necks? In every district one kind of animal ... alone could have its neck elongated.'

(below) 'Insects often resemble for the sake of protection various objects, such as leaves, dead twigs, lichen, flowers, or the excrement of birds. The resemblance is often wonderfully close, extending to form, and even to the manner in which the insects hold themselves. The caterpillars which project motionless like dead twigs from the bushes on which they feed offer an excellent instance.' **A caterpillar of the brimstone moth; it will remain camouflaged in this way throughout the day, only moving about to feed after nightfall.**

animals had their mental powers more highly developed than others, as such development would be advantageous to all? Why have not apes acquired the intellectual powers of man? Various causes could be assigned; but they are conjectural, and their relative probability cannot be weighed.

We will return to Mr. Mivart's other objections. Insects often resemble for the sake of protection various objects, such as leaves, dead twigs, lichen, flowers, or the excrement of birds. The resemblance is often wonderfully close, extending to form, and even to the manner in which insects hold themselves. The caterpillars which project motionless like dead twigs from the bushes on found in the stations frequented by them. Assuming that an insect happened to resemble in some degree a dead twig or decayed leaf, and that it varied slightly in many ways, then all the variations which rendered the insect more like any such object, and thus favoured its escape, would be preserved.

Nor can I see any force in Mr. Mivart's difficulty with respect to 'the last touches of perfection in the mimicry', as in the case given by Mr. Wallace of a walking-stick insect which resembles 'a stick grown over by a creeping moss or jungermannia'. So close was this resemblance that a native Dyak maintained that the foliaceous excrescences were really moss. Insects are preyed on by birds and other enemies, whose sight is probably sharper than ours, and every grade in resemblance which aided an insect to escape detection would tend towards its preservation.

The mammary glands are common to the whole class of mammals, and are indispensable for their existence. Mr. Mivart asks: 'Is it conceivable that the young of any animal was ever saved from destruction by accidentally sucking a drop of scarcely nutritious fluid from an accidentally hypertrophied cutaneous gland of its mother? And even if one was so, what chance was there of the perpetuation of such a variation?' But the case is not here put fairly. Mammals are descended from a marsupial form, and so the mammary glands will have been at first developed within the marsupial sack. In the case of the fish *Hippocampus* [sea-horse] the eggs are hatched and the young reared for a time within a sack of this nature; they are nourished by a secretion from the cutaneous glands of the sack. Now with the early progenitors of mammals, is it not possible that the young might have been similarly nourished? And in this case the individuals which secreted a fluid in some manner the most nutritious would in the long run have reared a larger number of well-nourished offspring than would the individuals which secreted a poorer fluid; and thus the cutan-

eous glands, which are the homologues of the mammary glands, would have been rendered more effective. It accords with the principle of specialisation that the glands over a certain space should have become more highly developed than the remainder; and they would then have formed a breast, but at first without a nipple, as we see in the Ornithorhyncus [duck-billed platypus] at the base of the mammalian series.

The development of the mammary glands would have been of no service unless the young were able to partake of the secretion. There is no greater difficulty in understanding how young mammals have instinctively learnt to suck the breast, than in understanding how unhatched chickens have learnt to break the egg-shell by tapping against it with their beaks. But the young kangaroo is said not to suck, only to cling to the nipple of its mother, who has the power of injecting milk into the mouth of her helpless, half-formed offspring. On this head Mr. Mivart remarks: 'Did no special provision exist, the young one must infallibly be choked by the intrusion of the milk into the windpipe. But there *is* a special provision. The larynx is so elongated that it rises up into the posterior end of the nasal passage, and is thus enabled to give free entrance to the air for the lungs, while the milk passes harmlessly on each side of this elongated larynx, and so safely attains the gullet behind it.' Mr. Mivart then asks how did natural selection remove in the adult kangaroo 'this at least perfectly innocent and harmless structure?' It may be suggested that the voice, which is of high importance to many animals, could hardly have been used with full force as long as the larynx entered the nasal passage; and this structure would have greatly interfered with an animal swallowing solid food.

In the vegetable kingdom Mr. Mivart alludes to the movements of climbing plants. These can be arranged in a long series, from those which simply twine round a support, to leaf-climbers, and to those provided with tendrils. In these two

A duck-billed platypus. The platypus and the spiny anteaters together form the Monotremes, a primitive group of mammals showing some features characteristic of reptiles. These animals have fur and are warm-blooded, but unlike other mammals they lay eggs. They produce milk but their mammary glands lack a nipple; when suckling the female platypus lies on her back and the young lap the milk up as it oozes out of simple openings in her belly. The Monotremes have survived only in Australasia.

points of the compass, one after the other in succession. By this movement the stems are made to move round and round. As soon as the lower part of a stem strikes any object and is stopped, the upper part still goes on revolving, and thus necessarily twines up the support. The revolving movement ceases after the early growth of each shoot. As, in many widely separated families, single species and genera have thus become twiners, they cannot have inherited it from a common progenitor. Hence some tendency to a movement of this kind cannot be uncommon with plants which do not climb; this has afforded the basis for natural selection to work on. Such is the case of the young flower-peduncles of a *Maurandia* which revolve slightly and irregularly, but without making any use of this habit. Similarly the young stems of an *Alisma* and a *Linum* revolve plainly though irregularly; and this occurs with some other plants. These slight movements are not of the least use in climbing. Nevertheless if it had profited them to ascend to a height, then the habit of irregularly revolving might have been increased through natural selection, until they had become well-developed twining species.

In ascending the series from simple twiners to leaf-climbers an important quality is added, namely sensitiveness to touch, by which the foot-stalks of the leaves or flowers, or these converted into tendrils, are excited to bend round and clasp the touching object. Nearly the same remarks are applicable as in the case of the revolving movements. As a vast number of species, belonging to widely distinct groups, are endowed with this sensitiveness, it ought to be found in a nascent condition in many plants which have not become climbers. This is the case: the young flower-peduncles of the above *Maurandia* curve themselves a little towards the side which is touched. In several species of *Oxalis* the leaves and their foot-stalks move, especially after exposure to a hot sun, when they are gently and repeatedly touched. The movement is distinct, but is best seen in the young

latter classes the stems have generally lost the power of twining. All the many gradations between simple twiners and tendril-bearers are beneficial in a high degree to the species.

As twining is the simplest means of ascending a support, it may naturally be asked how did plants acquire this power in an incipient degree. The power of twining depends firstly on the stems whilst young being extremely flexible (a character common to many plants which are not climbers), and secondly on their continually bending to all leaves; in others it is extremely slight. With climbing plants it is only during the early stages of growth that the foot-stalks and tendrils are sensitive.

I have now considered enough, perhaps more than enough, of the cases selected with care by a skilful naturalist, to prove that natural selection is incompetent to account for the incipient stages of useful structures; and I have shown, as I hope, that there is no great difficulty on this head.

Reasons for disbelieving in great and abrupt modifications

At the present day almost all naturalists admit evolution under some form. Mr. Mivart believes that species change through 'an internal force or tendency', about which it is not pretended that anything is known. That species have a capacity for change will be admitted by all evolutionists; but there is no need to invoke any internal force beyond the tendency to ordinary variability, which through natural selection would give rise by graduated steps to species. The final result will generally have been an advance, but in some few cases a retrogression, in organisation.

Mr. Mivart is further inclined to believe that new species manifest themselves 'with suddenness and by modifications appearing at once'. For instance he supposes that the differences between the extinct three-toed *Hipparion* and the horse arose suddenly. He thinks it difficult to believe that the wing of a bird 'was developed in any other way than by a comparatively sudden modification of a marked and important kind', and apparently he would extend the same view to the wings of bats and pterodactyls. This conclusion, which implies great breaks in the series, appears to me improbable in the highest degree.

That many species have been evolved in an extremely gradual manner, there can hardly be a doubt. The species and even the genera of many large families are so closely allied together that it is difficult to distinguish not a few of them. On

Many biologists failed to see how wings, such as those of birds and of bats, could have arisen except by a sudden major modification. Darwin maintained that they could have evolved gradually, the early forms of the wing being useful for gliding. Deductions from the fossil bird Archaeopteryx support this argument (*see p. 15*). Unfortunately no fossils have yet been found of animals ancestral to the bats.

every continent, in proceeding from north to south, from lowland to upland, etc., we meet with a host of closely related species.

Many large groups of facts are intelligible only on the principle that species have been evolved by very small steps. For instance, the fact that the species included in the larger genera are more closely related to each other, and present a greater number of varieties, than do species in the smaller genera, as was shown in our second chapter.

Although very many species have almost certainly been produced by gradual steps, yet it may be maintained that some have been developed in a different and abrupt manner. Such an admission, however, ought not to be made without strong evidence. One class of facts, namely the sudden appearance of new and distinct forms of life in our geological formations, supports at first sight the belief in abrupt development. But the value of this evidence depends entirely on the perfection of the geological record. If the record is as fragmentary as many geologists strenuously assert, there is nothing strange in new forms appearing as if suddenly developed.

Against the belief in such abrupt changes embryology enters a strong protest. It is notorious that the wings of birds and bats, and the legs of horses or other quadrupeds are indistinguishable at an early embryonic period, and that they become differentiated by fine steps. As we shall hereafter see, the embryo serves as a record of the past condition of the species. On this view, it is incredible that an animal should have undergone such momentous transformations as those above indicated, and yet should not bear even a trace in its embryonic condition of any sudden modification, every detail in its structure being developed by fine steps.

He who believes that some ancient form was transformed suddenly, through an internal force, into one furnished with wings, will be compelled to assume that many individuals varied simultaneously. It cannot be denied that such abrupt changes of structure are widely different from those which most species have undergone. He will further be compelled to believe that many structures beautifully adapted to all the other parts of the same creature, and to the surrounding conditions, have been suddenly produced; and of such complex and wonderful co-adaptations he will not be able to assign a shadow of an explanation. He will be forced to admit that these great transformations have left no trace of their action on the embryo. To admit all this is, as it seems to me, to enter into the realms of miracle, and to leave those of Science.

Chapter 8
Instinct

Many instincts are so wonderful that their development will probably appear to the reader a difficulty sufficient to overthrow my whole theory. I may here premise that I have nothing to do with the origin of the mental powers, any more than I have with that of life itself. We are concerned only with the diversities of instinct and other mental faculties in animals of the same class.

I will not attempt any definition of instinct, but everyone understands what is meant when it is said that instinct impels the cuckoo to migrate and to lay her eggs in other birds' nests. An action which we ourselves require experience to perform, when performed by an animal, especially a very young one without experience, and when performed by many individuals in the same way, without their knowing for what purpose, is usually said to be instinctive.

Frederick Cuvier and several of the older metaphysicians have compared instinct with habit. This comparison gives, I think, an accurate notion of the frame of mind under which an instinctive action is performed, but not necessarily of its origin. How unconsciously many habitual actions are performed! They easily become associated with other habits, with certain periods of time and states of the body, and when once acquired they often remain constant throughout life.

Several other points of resemblance between instincts and habits could be pointed out. As in repeating a well-known song, so in instincts, one action follows on another; if a person be interrupted in a song, or in repeating anything by rote, he is generally forced to go back to recover the habitual train of thought: so P. Huber found it

was with a caterpillar, which makes a very complicated hammock; for if he took a caterpillar which had completed its hammock up to the sixth stage of construction, and put it into a hammock completed up only to the third stage, the caterpillar simply reperformed the fourth, fifth, and sixth stages of construction. If, however, a caterpillar were taken out of a hammock made up to the third stage and put into one finished up to the sixth stage, far from deriving any benefit from this, it was much embarrassed, and in order to complete its hammock seemed forced to start from the third stage, where it had left off, and thus tried to complete the already finished work.

[*Habits, like other characteristics acquired in an individual's lifetime, cannot be passed on to its offspring. Therefore true instincts can only evolve by natural selection of spontaneous and random variations in already existing instincts. It should be mentioned, however, that in higher animals some behaviour which appears to be due to instinct actually depends on what is called 'cultural transmission'. For example, birds of many species are born with the instinct for their characteristic song pattern, but have to learn the details of the song by hearing others of the same species singing.*]

If we suppose any habitual action to be inherited – and it can be shown that this does sometimes happen – then the resemblance between what originally was a habit and an instinct becomes so close as not to be distinguished. If Mozart, instead of playing the pianoforte at three years old with wonderfully little practice, had played a tune with no practice at all, he might truly be said to have done so instinctively. But it would be a serious error to suppose that the greater number of instincts have been acquired by habit in one generation, and then transmitted by inheritance to succeeding generations. It can be clearly shown that the most wonderful instincts with which we are acquainted, namely those of the hive-bee and of many ants, could not possibly have been acquired by habit.

It will be universally admitted that instincts are as important as corporeal structures for the welfare of each species. Under changed conditions modifications of instinct might be profitable, and if it can be shown that instincts do vary ever so little, then I can see no difficulty in natural selection preserving and accumulating variations of instinct. As modifications of structure arise from habit and are lost by disuse, so I do not doubt it has been with instincts. But I believe that the effects of habit are in many cases of subordinate importance to the effects of the natural selection of spontaneous variations of instincts.

No instinct can be produced through natural selection except by the slow and gradual accumulation of numerous slight, yet profitable, variations. Hence, as in the case of corporeal structures, we ought to find in nature not the actual transitional gradations by which each complex instinct has been acquired – for these could be found only in lineal ancestors – but in the collateral lines of descent some evidence of such gradations; and this we do.

Again, as in the case of corporeal structure, the instinct of each species is good for itself but has never been produced for the exclusive good of others. One of the strongest instances of an animal apparently performing an action for the sole good of another is that of aphids voluntarily yielding their sweet excretion to ants. I removed all the ants from a group of aphids on a dock-plant, and prevented their attendance during several hours. After this interval I felt sure that the aphids would want to excrete, but not one did; I then tickled and stroked them with a hair in the same manner, as well as I could, as the ants do with their antennae;

but not one excreted. Afterwards I allowed an ant to visit them; each aphid, as soon as it felt the ant's antennae playing on its abdomen, immediately excreted a limpid drop of sweet juice, which was eagerly devoured by the ant. Even the young aphids behaved in this manner, showing that the action was instinctive and not the result of experience. If ants be not present the aphids are at last compelled to eject their excretion. However, as the excretion is extremely viscid, it is no doubt a convenience to the aphids to have it removed; therefore probably they do not excrete solely for the good of the ants. Although no animal performs an action for the exclusive good of another species, yet each tries to take advantage of the instincts of others, as each takes advantage of the weaker bodily structure of other species. So again certain instincts cannot be considered as absolutely perfect.

Instincts certainly do vary – for instance the migratory instinct, both in extent and direction, and in its total loss. So it is with the nests of birds; Audubon has given cases of differences in the nests of the same species in the northern and southern United States. Fear of any particular enemy is certainly an instinctive quality, as in nestling birds, though it is strengthened by experience and the sight of fear of the same enemy in other animals. The fear of man is slowly acquired by the various animals which inhabit desert islands; and we see an instance of this in England in the great wildness of our large birds in comparison with our small birds, for they have been most persecuted by man. We may safely attribute the greater wildness of our large birds to this cause, for in uninhabited islands large birds are not more fearful than small; the magpie, so wary in England, is tame in Norway, as is the hooded crow in Egypt.

Inherited Changes of Habit or Instinct in Domesticated Animals

The probability of inherited variations of instinct in a state of nature will be strengthened by briefly

considering a few cases under domestication. It is notorious how much domestic animals vary in their mental qualities. With the breeds of the dog, young pointers will sometimes point and even back other dogs the very first time that they are taken out; retrieving is in some degree inherited by retrievers, and so is a tendency to run round, instead of at, a flock of sheep by shepherd-dogs.

How strongly these domestic instincts are inherited and how curiously they mingle is well shown when different breeds of dogs are crossed. Thus a cross with a bulldog affects for many generations the courage and obstinacy of greyhounds; and a cross with a greyhound has given to a whole family of shepherd-dogs a tendency to hunt hares.

Domestic instincts are sometimes spoken of as actions which have become inherited solely from long-continued habit, but this is not true. No one would have thought of training a dog to point had not some dog naturally shown this tendency, and this is known occasionally to happen, as I once saw, in a pure terrier: the act of pointing is probably only the exaggerated pause of an animal preparing to spring on its prey.

Hence we may conclude that under domestication instincts have been acquired and natural instincts have been lost by man selecting and accumulating, during successive generations,

Ants 'milking' aphids for drops of honeydew (*above*). The excretion of honeydew by aphids is simply a by-product of their manner of feeding: since the plant sap which they suck is high in carbohydrates but low in protein, large quantities of it must be ingested to obtain enough protein, and the excess carbohydrate excreted as honeydew. Darwin observed that the aphids readily cooperated with the ants, and, if no ants were present, saved up their honeydew until an ant returned. He was puzzled by this since he was firmly convinced that no instinct could be produced *'for the exclusive good of another species'*, and suggested that it must be of advantage to the aphids to have the sticky excretion removed. More recent studies have shown that the aphids can flick the honeydew away from themselves if ants are not in attendance, and that the main benefit aphids derive from their relationship with ants is protection from predators. Many insects, such as the hoverfly larva (*below*), eat aphids, but the presence of ants, which both bite and emit acid in self-defence, will deter them.

peculiar mental habits and actions which at first appeared from what we must in our ignorance call an accident.

Special Instincts

We shall, perhaps, best understand how instincts in a state of nature have become modified by selection by considering three cases: the instinct which leads the cuckoo to lay her eggs in other birds' nests, the slave-making instinct of certain ants, and the cell-making power of the hive-bee.

Instinct of the Cuckoo

It is supposed by some naturalists that the immediate cause of the instinct of the cuckoo is that she lays her eggs at intervals of two or three days; so that, if she were to sit on her own eggs, those first laid would have to be left for some time unincubated, or there would be eggs and young birds of different ages in the same nest. If this were the case, the process of laying and hatching might be inconveniently long, especially as she migrates at a very early period. The American cuckoo is in this predicament; for she makes her own nest, and has eggs and young successively hatched, all at the same time.

It has been both asserted and denied that the American cuckoo occasionally lays her eggs in other birds' nests, and I could give several instances of various birds which have been known to occasionally do this. Now let us suppose that the ancient progenitor of our European cuckoo had the habits of the American cuckoo, and that she occasionally laid an egg in another birds' nest. If the old bird profited through being enabled to migrate earlier, or if the young were made more vigorous than when reared by their own mother, encumbered as she could hardly fail to be by having eggs and young of different ages at the same time, then the old birds or the fostered young would gain an advantage. And the young thus reared would be apt to follow by inheritance the occasional habit of their mother, and in turn would be apt to lay their eggs in other birds' nests, and thus be more successful in rearing their young. By a continued process of this nature, I believe that the strange instinct of our cuckoo has been generated. It has also recently been ascertained that the cuckoo occasionally lays her eggs on the bare ground, sits on them, and feeds her young. This rare event is probably a case of reversion to the long-lost, aboriginal instinct.

In the case of the European cuckoo, the chief points to be noted are three: first, that the cuckoo lays only one egg in a nest, so that the voracious young bird receives ample food. Secondly, that

Darwin noted that, whereas in Britain large birds were much less tame than small ones, this was not so on oceanic islands. Of the Galapagos hawk he wrote in *The Zoology of the Voyage of H.M.S. Beagle*: *'It is extremely tame and frequents the neighbourhood of any building inhabited by man. When a tortoise is killed, even in the midst of woods, these birds congregate in great numbers, and remain either seated on the ground, or on the branches of the stunted trees, patiently waiting to devour the intestines ...'* He attributed the wildness of large birds in Britain to the fact that they had been *'most persecuted by man'*.

A pointer dog ranging over open ground in advance of huntsmen. When its sensitive nose picks up the scent of game it will freeze, and stand motionless with its nose pointing in the direction of the game. This characteristic behaviour of the breed is instinctive.

the eggs are remarkably small: a real case of adaptation, as the non-parasitic American cuckoo lays full-sized eggs. Thirdly, that the young cuckoo, soon after birth, has the instinct, the strength and a properly shaped back for ejecting its foster-brothers, which then perish from cold and hunger.

This strange and odious instinct was probably acquired so that the young cuckoo should receive as much food as possible. I see no special difficulty in its having gradually acquired, during successive generations, the blind desire, strength, and structure necessary for the work of ejection; for those young cuckoos which had such habits and structure best developed would be the most securely reared. The first step towards the acquisition of the proper instinct might have been mere unintentional restlessness on the part of the young bird, when somewhat advanced in age and strength; the habit having been afterwards improved and transmitted to an earlier age. For if each part is liable to individual variations at all ages, and the variations tend to be inherited at a corresponding age – propositions which cannot be disputed – then the instincts and structure of the young

could be slowly modified as surely as those of the adult.

Some species of *Molothrus*, a distinct genus of American birds allied to our starlings, have parasitic habits like those of the cuckoo; and the species present an interesting gradation in the perfection of their instincts. Birds of the species *Molothrus badius*, [bayheaded or baywinged cowbird], either build a nest of their own or seize on one belonging to some other bird, occasionally throwing out the nestlings. They either lay their eggs in the nest thus appropriated, or oddly enough build one for themselves on top of it. They usually sit on their own eggs and rear their own young; but that excellent observer Mr. Hudson says it is probable that they are occasionally parasitic, for he has seen the young of this species following old birds of a distinct kind and clamouring to be fed by them.

The parasitic habits of another species, *Molothrus bonariensis*, [shiny cowbird], are much more highly developed, but are still far from perfect. This bird invariably lays its eggs in the nests of strangers; but it is remarkable that several together sometimes commence to build an irregular untidy nest of their own, placed in singularly ill-adapted situations, as on the leaves of a large thistle. They never, however, complete a nest for themselves. They often lay so many eggs – from fifteen to twenty – in the same foster-nest, that few can possibly be hatched. They have, moreover, the extraordinary habit of pecking holes in the eggs, whether of their own species or of their foster-parents, which they find in the appropriated nests.

A third species, *Molothrus pecoris*, [brown-headed cowbird] of North America, has acquired instincts as perfect as those of the cuckoo, for it never lays more than one egg in a foster-nest, so that the young bird is securely reared. Mr. Hudson is a strong disbeliever in evolution, but he appears to have been so much struck by the imperfect instincts of *Molothrus bonariensis* that he quotes my words and asks, 'Must we consider these habits, not as especially endowed or created instincts, but as small consequences of one general law, namely transition?'.

Many bees are parasitic, and regularly lay their eggs in the nests of other kinds of bees. These bees have not only had their instincts but their structure modified in accordance with their parasitic habits, for they do not possess the pollen-collecting apparatus which would have been indispensable if they had stored up food for their own young.

Some species of Sphegidae, [sandwasps and digger-wasps], are likewise parasitic; and M. Fabre has lately shown that although the species *Tachytes nigra* generally makes its own burrow and stores it with paralysed prey for its larvae, yet when it finds a burrow already made and stored by another, it takes advantage of the prize and becomes, for the occasion, parasitic. In this case, as with that of the *Molothrus* or cuckoo, I can see no difficulty in natural selection making an occasional habit permanent if of advantage to the species, and if the insect whose nest and stored food are appropriated be not thus exterminated.

Slave-making instinct

This remarkable instinct was first discovered in *Formica rufescens*. This ant is absolutely dependent on its slaves. The males and fertile females do no work of any kind, and the workers or sterile females, though most energetic in capturing slaves, do no other work. They are incapable of making their own nests or feeding their own larvae. When they have to migrate it is the slaves which determine the migration and actually carry their masters in their jaws. So utterly helpless are the masters, that when Pierre Huber shut up thirty of them without a slave but with plenty of food and with their own larvae and pupae to stimulate them to work, they did nothing; they could not even feed themselves, and many perished of hunger. Huber then introduced a single slave (*F. fusca*), and she instantly set to work, fed and saved the

survivors, made some cells, tended the larvae, and put all to rights. If we had not known of any other slave-making ant, it would have been hopeless to speculate how so wonderful an instinct could have been perfected.

Another species, *Formica sanguinea*, is likewise a slave-making ant. I opened fourteen nests of *F. sanguinea* and found a few slaves in all, these being sterile females of the species *F. fusca*; males and fertile females of the species are found only in their own proper communities, and have never been observed in the nests of *F. sanguinea*. The slaves are black and not above half the size of their red masters, so that the contrast in their appearance is great. When the nest is slightly disturbed the slaves occasionally come out, and like their masters are much agitated and defend the nest. Hence it is clear that the slaves feel quite at home. On three successive years I watched for many hours several nests in Surrey and Sussex, and never saw a slave either leave or enter a nest. The masters, on the other hand, constantly bring in materials for the nest and food of all kinds. During the year 1860, however, I came across a community with an unusually large stock of slaves, and I observed a few slaves mingled with their masters leaving the nest and marching to a tall tree which they ascended together, probably in search of aphids. According to Huber, the slaves in Switzerland habitually work with their masters in making the nest; they alone open and close the doors in the morning and evening, and their principal office is to search for aphids. This difference in the usual habits of the masters and slaves in the two countries probably depends merely on the slaves being captured in greater numbers in Switzerland than in England.

One day I witnessed a migration of *F. sanguinea* from one nest to another, and it was most interesting to behold the masters carefully carrying their slaves in their jaws instead of being carried by them, as in the case of *F. rufescens*. Another day my attention was struck by about a score of the slave-makers haunting the same spot, and evidently not

A newly hatched cuckoo moves awkwardly about the nest, ejecting every other object it encounters. A young nestling has already been pushed out onto the rim of the nest. The cuckoo chick has now come up against an egg, and is pushing it with its throat, attempting to squeeze the egg upwards, over its head and into the hollow of its back. This hollow has the specific function of ejecting eggs: once the cuckoo has the egg in the hollow it can lift itself up and tip the egg out of the nest. Darwin saw that the physical adaptations and extraordinary instincts of the cuckoo chick could have evolved by natural selection: *'This strange and odious instinct was probably acquired so that the young cuckoo should receive as much food as possible. I see no special difficulty in its having gradually acquired, during successive generations, the blind desire, strength, and structure necessary for the work of ejection; for those young cuckoos which had such habits and structure best developed would be the most securely reared. The first step towards the acquisition of the proper instinct might have been mere unintentional restlessness on the part of the young bird, when somewhat advanced in age and strength; the habit having been afterwards improved and transmitted to an earlier age.'*

The brown-headed cowbird of North America shows similar behaviour to that of the cuckoo, laying only one egg in each foster nest. Darwin argues that the strange behaviour patterns of other species of cowbirds should be seen as transitional stages between normal (i.e. non-parasitic) nesting behaviour and the perfectly developed parasitism of the brown-headed cowbird.

Slave-making ants, *Formica sanguinea,* attacking workers of the species *Formica fusca,* and carrying off their pupae to be reared as slaves. From a photograph by H. Bastin.

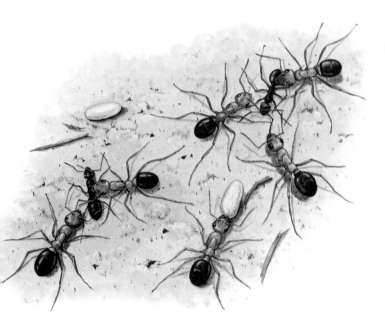

in search of food; they approached and were vigorously repulsed by an independent community of the slave-species (*F. fusca*). The slave-makers killed their small opponents and carried their dead bodies as food to their nest, but they were prevented from getting any pupae to rear as slaves. I then put a small parcel of the pupae of *F. fusca* down near the place of combat; they were eagerly seized and carried off by the tyrants, who perhaps fancied that, after all, they had been victorious in their combat.

At the same time I laid on the same place a small parcel of the pupae of another species, *F. flava,* with a few of these little yellow ants still clinging to the fragments of their nest. This species is sometimes, though rarely, made into slaves. Although so small a species, it is very courageous, and I have seen it ferociously attack other ants. I was curious to ascertain whether *F. sanguinea* could distinguish the pupae of *F. fusca,* which they habitually make into slaves, from those of the little and furious *F. flava,* and it was evident that they did so, for they were much terrified when they came across the pupae, or even the earth from the nest, of *F. flava,* and quickly ran away; but in about a quarter of an hour, shortly after all the little yellow ants had crawled away, they took heart and carried off the pupae.

One evening I visited another community of *F. sanguinea,* and found a number of these ants returning home carrying dead bodies of *F. fusca* and numerous pupae. I traced a long file of ants burdened with booty back to a very thick clump of heath, whence I saw the last individual of *F. sanguinea* emerge, carrying a pupa; but I was not able to find the desolated nest in the thick heath. The nest, however, must have been close at hand, for two or three individuals of *F. fusca* were rushing about in the greatest agitation, and one was perched motionless with its own pupa in its mouth on the top of a spray of heather, an image of despair over its ravaged home.

Such are the facts in regard to the instinct of

making slaves. Let it be observed what a contrast the habits of *F. sanguinea* present with those of *F. rufescens*. The latter does not build its own nest, does not determine its own migrations, does not collect food, and cannot even feed itself: it is absolutely dependent on its numerous slaves. *Formica sanguinea*, on the other hand, possesses much fewer slaves: the masters determine when and where new nests shall be formed, and when they migrate the masters carry the slaves. The slaves have the exclusive care of the larvae, and the masters alone go on slave-making expeditions. In Switzerland the slaves and masters work together, making and bringing materials for the nest; both, but chiefly the slaves, tend and milk their aphids. In England the masters alone usually leave the nest to collect building materials and food, so that here they receive much less service from their slaves.

By what steps the instinct of *F. sanguinea* originated I will not pretend to conjecture. But as ants which are not slave-makers will carry off the pupae of other species, it is possible that such pupae originally stored as food might become developed; and the foreign ants thus unintentionally reared would then follow their proper instincts, and do what work they could. If their presence proved useful to the species which had seized them – if it were more advantageous to this species to capture workers than to procreate them – the habit of collecting pupae, originally for food, might by natural selection be strengthened and rendered permanent for the very different purpose of raising slaves. When the instinct was once acquired natural selection might increase and modify the instinct – always supposing each modification to be of use to the species – until an ant was formed as abjectly dependent on its slaves as is the *Formica rufescens*.

Cell-making instinct of the Hive-Bee

He must be a dull man who can examine the exquisite structure of a comb without enthusiastic admiration. We hear from mathematicians that bees have made their cells of the proper shape to hold the greatest amount of honey, with the least consumption of precious wax in their construction. A skilful workman with fitting tools and measures would find it very difficult to make cells of wax of the true form, though this is effected by a crowd of bees working in a dark hive. It seems inconceivable how they can make all the necessary angles and planes, or even perceive when they are correctly made. But the difficulty is not nearly so great as it first appears: all this beautiful work can be shown to follow from a few simple instincts.

Let us look to the great principle of gradation, and see whether Nature does not reveal to us her method of work. At one end of a short series we have humble-bees [bumble-bees], which use their old cocoons to hold honey, sometimes adding to them short tubes of wax and making separate and very irregular rounded cells of wax. At the other end of the series we have the cells of the hive-bee, placed in a double layer, each cell being a hexagonal prism. In the series between the cells of the hive-bee and the humble-bee we have the cells of the Mexican *Melipona domestica*. The *Melipona* itself is intermediate in structure between the hive- and humble-bee, but more nearly related to the latter; it forms a nearly regular waxen comb of cylindrical cells, in which the young are hatched, and some large cells of wax for holding honey. These latter cells are nearly spherical and of nearly equal sizes, and are aggregated into an irregular mass. But the important point is that these cells are always made so near to each other that they would have broken in to each other if the spheres had been completed; but this is never permitted, the bees building perfectly flat walls of wax between the spheres which thus tend to intersect. Hence, each cell consists of an outer spherical portion and of two, three, or more flat surfaces, according as the cell adjoins two, three, or more other cells. As in the cells of the hive-bee, so here the three plane surfaces in any one cell necessarily enter into the

Bumble-bees (*above*) construct rounded wax cells of irregular size for the storage of honey, while the cells of the honey-bee (*below*) are perfect hexagons of identical size. Darwin shows how the instincts necessary to construct such perfect cells could have evolved from simpler instincts like those of the bumble-bee.

construction of three adjoining cells. It is obvious that the *Melipona* saves wax and labour by this manner of building.

Reflecting on this case, it occurred to me that if the *Melipona* had made its spheres at some given distance from each other, and had made them of equal sizes and arranged them symmetrically in a double layer, the resulting structure would have been as perfect as the comb of the hive-bee. Accordingly I wrote to Professor Miller of Cambridge, and this geometer tells me that it is strictly correct.

Hence if we could slightly modify the instincts already possessed by the *Melipona*, this bee would make a structure as perfect as that of the hive-bee. We must suppose the *Melipona* to have the power of forming her cells truly spherical and of equal sizes; and this would not be very surprising, seeing that she already does so to a certain extent. We must suppose the *Melipona* to arrange her cells in level layers, as she already does her cylindrical cells; and we must further suppose that she can somehow judge accurately at what distance to stand from her fellow-labourers when several are making their spheres; but she is already so far enabled to judge distance that she always describes her spheres to intersect to a certain extent, and then unites the points of intersection by flat surfaces. By such modifications of instincts which in themselves are not very wonderful – hardly more wonderful than those which guide a bird to make its nest – I believe that the hive-bee has acquired through natural selection her inimitable architectural powers.

Objections to the Theory of Natural Selection as applied to Instincts: Neuter and Sterile Insects

It has been objected to the foregoing view of the origin of instincts that 'the variations of structure and of instinct must have been simultaneous . . . as a modification in the one without an immediate corresponding change in the other would have been fatal.' This objection rests on the assumption that the changes in the instincts and structure are abrupt. Take as an illustration the case of the larger titmouse [great tit], *Parus major*, alluded to in a previous chapter; this bird often holds the seeds of the yew between its feet on a branch, and hammers with its beak till it gets at the kernel. Now what special difficulty would there be in natural selection preserving all the slight variations in the shape of the beak which were better adapted to break open the seeds, until a beak was formed as well constructed for this purpose as that of the nuthatch, at the same time that habit or spontaneous variations of taste led the bird to become more and more of a seed-eater?

To take another case: few instincts are more remarkable than that which leads the swift of the Eastern Islands to make its nest wholly of saliva. Some birds build their nests of mud moistened with saliva; and one of the swifts of North America makes its nest of sticks agglutinated with saliva, and even with flakes of this substance. Is it then improbable that natural selection of individual swifts which secreted more and more saliva should at last produce a species with instincts leading it to make its nest exclusively of saliva?

No doubt many instincts could be opposed to the theory of natural selection, but I will confine myself to one special difficulty: the neuters or sterile females in insect-communities. For these neuters often differ widely from both males and fertile females, and yet, from being sterile, they cannot propagate their kind.

I will here take only the case of working or sterile ants. How the workers have been rendered sterile is a difficulty; but some insects and other animals in a state of nature do occasionally become sterile, and if such insects had been social, and it had been profitable to the community that a number should have been born capable of work, but incapable of procreation, I can see no especial difficulty in this having been effected through natural selection. The greater difficulty lies in the working ants differing widely from both males and fertile females in structure, as in the shape of the thorax, and in being destitute of wings and sometimes of eyes, and in instinct. As far as instinct alone is concerned, the wonderful difference between workers and fertile females would have been better exemplified by the hive-bee. If a working ant had been an ordinary animal, I should have unhesitatingly assumed that all its characters had been slowly acquired through the inheritance of profitable modifications. But with the working ant we have an insect differing greatly from its parents, yet absolutely sterile; so it could never have transmitted modifications to its progeny. It may well be asked how is it possible to reconcile this case with the theory of natural selection?

First, let us remember the innumerable instances of differences of inherited structure which are correlated with certain ages, and with either sex. We have differences correlated not only with one sex but with that short period when the reproductive system is active, as in the nuptial plumage of many birds and the hooked jaws of the male salmon. Hence I see no difficulty in any character becoming correlated with the sterile condition of certain members of insect-communities: the difficulty lies in how such correlated modifications could have been accumulated by natural selection.

This difficulty disappears when it is remembered that selection may be applied to the family as well as to the individual, and may thus gain the desired end. If slight modifications of structure or instinct, correlated with the sterile condition of certain members of the community, prove advantageous, then the fertile males and females will transmit to their fertile offspring a tendency to produce sterile members with the same modifications. This process must have been repeated many times, until that prodigious amount of difference between the fertile and sterile females of the same species has been produced which we see in many social insects.

But we have not as yet touched on the acme of the difficulty, namely the fact that the neuters of several ants differ not only from the fertile females and males but from each other, and are thus divided into two or even three castes. The castes, moreover, do not commonly graduate into each other, but are perfectly well defined. Thus in *Eciton* there are working and soldier neuters, with jaws and instincts extraordinarily different. In the simpler case of neuter insects all of one caste, we may conclude from the analogy of ordinary variations that the successive modifications did not first arise in all the neuters in the same nest, but in some few alone, and that by the survival of the com-

munities with females which produced most neuters having the advantageous modification, all the neuters ultimately came to be thus charac- terised. Thus we ought occasionally to find neuter insects presenting gradations of structure; and this we do, for the extreme forms of the neuters of several British ants can be linked together by individuals taken out of the same nest: I have my- self compared perfect gradations of this kind amongst workers of *Formica flava*.

I also examined numerous neuters from the same nest of the driver ant of West Africa. The reader will best appreciate the difference in these workers by an illustration: the difference was the same as if we were to see workmen building a house, of whom many were five feet four inches high, and many sixteen feet high; in addition

Three different types of worker ants of a Brazilian leaf-cutting species, showing the enormous differences in size and structure which Darwin discusses.

Saüba or Leaf-carrying Ant.—1. Worker-minor; 2. Worker-major; 3. Subterranean worker.

suppose that the larger workmen had heads four times as big as those of the smaller men, and jaws nearly five times as big. The jaws, moreover, of the working ants of the several sizes differed wonderfully in shape and in the form and number of the teeth. But the important fact is that, though the workers can be grouped into castes of different sizes, yet they graduate insensibly into each other, as does the widely different structure of their jaws.

With these facts before me, I believe that natural selection could form a species which should regularly produce one set of workers of one size and structure, and simultaneously another set of a different size and structure, a graduated series having first been formed, and then the extreme forms having been produced in greater numbers through the survival of the parents which generated them, until none with an intermediate structure were produced.

I have discussed this case to show the power of natural selection, and because this is by far the most serious special difficulty which my theory has encountered. The case, also, is very interesting, as it proves that with animals, as with plants, any amount of modification may be effected by the accumulation of spontaneous profitable variations, without exercise or habit having been brought into play. For peculiar habits confined to the workers could not possibly affect the males and fertile females, which alone leave descendants. I am surprised that no one has hitherto advanced this demonstrative case of neuter insects against the well-known doctrine of inherited habit, as advanced by Lamarck.

Summary

I have endeavoured in this chapter briefly to show that the mental qualities of our domestic animals vary, and that the variations are inherited. Still more briefly I have attempted to show that instincts vary slightly in a state of nature. No one will dispute that instincts are of the highest importance to each animal. Therefore there is no real difficulty, under changing conditions of life, in natural selection accumulating to any extent slight modifications of instinct which are in any way useful.

I do not pretend that the facts given in this chapter strengthen in any great degree my theory; but none of the cases of difficulty, to the best of my judgment, annihilate it. On the other hand, the fact that instincts are not always absolutely perfect and are liable to mistakes, that no instinct can be shown to have been produced for the good of other animals though animals take advantage of the instincts of others, that the canon in natural history of *Natura non facit saltum* is applicable to instincts as well as to corporeal structure, and is plainly explicable on the foregoing views, but is otherwise inexplicable – all tend to corroborate the theory of natural selection.

The theory of natural selection is also strengthened by some few other facts in regard to instincts; as by that common case of closely allied but distinct species, when inhabiting distant parts of the world and living under considerably different conditions of life, yet often retaining nearly the same instincts. For instance, we can understand, on the principle of inheritance, how it is that the thrush of tropical South America lines its nest with mud in the same peculiar manner as does our British thrush; and how it is that the Hornbills of Africa and India have the same extraordinary instinct of plastering up and imprisoning the females in a hole in a tree, with only a small hole left in the plaster through which the males feed them and their young when hatched.

Finally, it may not be a logical deduction, but to my imagination it is far more satisfactory to look at such instincts as the young cuckoo ejecting its foster-brothers, ants making slaves, or the larvae of ichneumonidae feeding within the live bodies of caterpillars, not as specially endowed or created instincts but as small consequences of one general law leading to the advancement of all organic beings, namely: multiply, vary, let the strongest live and the weakest die.

Ichneumon flies lay their eggs inside a caterpillar or its pupa. When the eggs hatch the ichneumon larvae eat the caterpillar alive, then metamorphose into adult ichneumon flies. The picture shows a young ichneumon fly emerging from the chrysalis of a butterfly. Darwin found it less distressing to believe that such cruel instincts as these had evolved by natural selection, than that they had been created by God. He wrote to Asa Gray in a letter of 1860: *'There seems to me too much misery in the world. I cannot persuade myself that a beneficient and omnipotent God would have designedly created the Ichneumonidae with the express intention of their feeding within the living bodies of caterpillars . . .'*

Chapter 9
Hybridism

[This chapter continues the argument that species are merely 'well-marked varieties'. A modern biologist would define a species as a group all of which can potentially interbreed one with another. In other words, it is a group within which genetic material can flow freely, but which is genetically isolated from other groups. The species can be recognized as an especially meaningful category in taxonomy because its genetic isolation permits it to evolve independently, thus producing distinctive features. However the whole natural world is not neatly divided into species, and the line between sterility and fertility is often blurred, as, for example, when two species are still in the process of separating from each other. With some species of plants there is sterility between particular individuals, although these quite clearly belong to the same species.

In Darwin's time a species was seen as a distinct group created by God. The taxonomist's job was merely to decide where the division between these created species lay, and the criterion of fertility within the group was an important one in coming to this decision. But where species were not clear-cut, such an approach encountered great difficulties. In this chapter Darwin emphasizes these difficulties in order to show the Creationist view of species to be untenable.]

The view commonly entertained by naturalists is that species, when intercrossed, have been specially endowed with sterility in order to prevent their confusion. This view certainly seems at first highly probable, for species living together could hardly have been kept distinct had they been capable of freely crossing.

In treating this subject, two classes of facts, to a large extent fundamentally different, have generally been confounded; namely, the sterility of species when first crossed, and the sterility of the hybrids produced from them. Pure species have, of course, their organs of reproduction in a perfect condition, yet when intercrossed they produce either few or no offspring. Hybrids, on the other hand, have their reproductive organs functionally impotent, as may be clearly seen in the state of the male element in both plants and animals.

As regards the sterility of species when crossed and of their hybrid offspring, it is impossible to study the several memoirs and works of those two conscientious and admirable observers, Kölreuter and Gärtner, who almost devoted their lives to this subject, without being deeply impressed with the high generality of some degree of sterility. But the sterility of various species when crossed is so different in degree and graduates away so insensibly, and, on the other hand, the fertility of pure species is so easily affected by various circumstances, that for all practical purposes it is most difficult to say where perfect fertility ends and sterility begins.

It is surprising in how many curious ways this gradation can be shown. When pollen from a plant of one family is placed on the stigma of a plant of a distinct family, it exerts no more influence than so much inorganic dust. From this absolute zero of fertility, the pollen of different species applied to the stigma of some one species of the same genus yields a perfect gradation in the number of seeds produced, up to nearly complete or even quite complete fertility; and, in certain abnormal cases, even to an excess of fertility, beyond that which the plant's own pollen produces.

So in hybrids between species, there are some

which never have produced, and probably never would produce, even with the pollen of the pure parents, a single fertile seed: but in some of these cases a first trace of fertility may be detected, by the pollen of one of the pure parent-species causing the flower of the hybrid to wither earlier than it otherwise would have done; and the early withering of the flower is well known to be a sign of incipient fertilisation. From this extreme degree of sterility we have self-fertilised hybrids producing a greater and greater number of seeds up to perfect fertility.

Although I know of hardly any thoroughly well-authenticated cases of perfectly fertile hybrid animals, I have reason to believe that the hybrids from *Phasianus colchicus* [ring-necked pheasant] with *P. torquatus* [now regarded as a subspecies of the former], are perfectly fertile. M. Quatrefages states that the hybrids from two moths (*Bombyx cynthia* and *arrindia*) were proved in Paris to be fertile *inter se* for eight generations. The hybrids from the common and Chinese geese, species which are so different that they are generally ranked in distinct genera, have often bred in this country with either pure parent, and in one single instance they have bred *inter se*.

It may be concluded that some degree of sterility, both in first crosses and in hybrids, is an extremely general result; but that it cannot be considered as absolutely universal. Considering the rules which govern the fertility of first crosses and of hybrids, we see that their fertility is eminently susceptible to favourable and unfavourable conditions, and is innately variable. It is by no means always the same in degree in the first cross and in the hybrids produced from this cross, and the fertility of hybrids is not related to the degree in which they resemble in external appearance either parent. Lastly, the facility of making a first cross between any two species is not always governed by their systematic affinity or degree of resemblance to each other. This latter statement is clearly proved by the difference in the result of reciprocal crosses between the same two species, for, according as the one species or the other is used as the father or the mother, there is generally some difference, and occasionally the widest possible difference, in the facility of effecting a union. The hybrids, moreover, produced from reciprocal crosses often differ in fertility.

The reproductive organs of a tulip flower. The male organs are known as anthers and produce powdery pollen. In this tulip both anthers and pollen are blackish in colour, although in most plants they are yellow. The female organ, which in this flower is white, has three parts: the four-lobed stigma which is sticky and receives the pollen; the fleshy stalk that supports it known as the style; and, at the base of the style, the ovaries *(not visible in this picture)*. For fertilization to occur the pollen must grow down through the style to reach the ovaries, forming a pollen-tube. Pollen is normally carried from one flower to another by insects, except in wind-pollinated flowers. Darwin grew various plants and crossed them artificically, transferring the pollen with a paintbrush, to see whether they would hybridize.

Now do these complex and singular rules indicate that species have been endowed with sterility simply to prevent their becoming confounded in nature? I think not. For why should the sterility be so extremely different in degree when various species are crossed, all of which we must suppose it would be equally important to keep from blending together? Why should some species cross with facility, and yet produce very sterile hybrids; and other species cross with extreme difficulty, and yet produce fairly fertile hybrids? Why should there often be so great a difference in the result of a reciprocal cross between the same two species? Why, it may even be asked, has the production of hybrids been permitted? To grant to species the special power of producing hybrids, and then to stop their further propagation by different degrees of sterility, seems a strange arrangement.

The foregoing rules and facts, on the other hand, appear to me clearly to indicate that the sterility of first crosses and of hybrids is simply incidental to unknown differences in their reproductive systems; the differences being of so peculiar and limited a nature that in reciprocal crosses between the same two species the male sexual element of the one will often freely act on the female sexual element of the other, but not in a reversed direction.

Origin and Causes of the Sterility of First Crosses and of Hybrids

[*The differences which prevent breeding between species are now called by the general term 'isolating mechanisms'. These include differences in courtship behaviour and breeding season, as well as physical differences which prevent mating or development of the embryo. Geographical isolation, i.e. isolation by distance or by physical barriers such as water or mountains, can also be an important isolating mechanism.*]
We will now look a little closer at the probable nature of the differences between species which induce sterility in first crosses and in hybrids. In the case of first crosses, the difficulty in effecting a union and obtaining offspring apparently depends on several distinct causes. There must sometimes be a physical impossibility in the male element reaching the ovule, as would be the case with a plant having a pistil too long for the pollen-tubes to reach the ovarium. It has also been observed that when the pollen of one species is placed on the stigma of a distantly allied species, though the pollen-tubes protrude, they do not penetrate the stigmatic surface. Again, the male element may reach the female element but be incapable of causing an embryo to be developed. Lastly, an embryo may be developed and then perish at an early period. With plants it is also known that hybrids raised from very distinct species are sometimes weak and dwarfed, and perish at an early age.

In regard to the sterility of hybrids, in which the sexual elements are imperfectly developed, the case is somewhat different. The sterility is independent of general health, and is often accompanied by excess of size or great luxuriance. The sterility occurs in various degrees; the male element is the most liable to be affected; but sometimes the female more than the male, and the tendency goes to a certain extent with systematic affinity, for whole groups of species tend to produce sterile hybrids. On the other hand certain species in a group will produce unusually fertile hybrids. No one can tell till he tries whether any two species of a genus will produce sterile hybrids.

Thus we see that when hybrids are produced by the unnatural crossing of two species, the reproductive system, independently of the general state of health, is affected by two distinct structures and constitutions having been blended into one. For it is scarcely possible that two organisations should be compounded into one, without some disturbance occurring in the development or mutual relations of the different parts and organs one to another. It must, however, be owned that we cannot understand on the above view several

Crosses are possible between all the species of the genus *Equus:* horses, donkeys and zebras. The best known hybrid is that from the horse x donkey cross, the mule, noted for its strength and stamina. The hybrids between horses and zebras, and between donkeys and zebras are also healthy and vigorous, but like the mule are sterile. The picture above shows horse x zebra hybrids, or zebroids, galloping with horses. The picture on the left shows a donkey with a foal fathered by a zebra. Darwin asks of those who believe that the Creator has endowed species with sterility to prevent their becoming confused: '*Why . . . has the production of hybrids been permitted? To grant to species the special power of producing hybrids, and then to stop their further propagation by different degrees of sterility, seems a strange arrangement.*'

facts with respect to the sterility of hybrids, for instance the unequal fertility of hybrids produced from reciprocal crosses.

[*It is now known that there can be natural selection for isolating mechanisms. Where, for example, two species occupy different habitats but are still able to interbreed, the hybrid between them may be unable to survive in either habitat. Any inherited mechanism which reduces crossing between individuals of the two species will be favoured by natural selection, since it will be advantageous not to waste reproductive effort in producing maladapted hybrids. In some cases the chromosomes of the different species cannot pair properly during meiosis (see diagram on p. 24), because, in the course of time, they have developed structural differences. The hybrid can still be healthy and vigorous but it will be sterile or have reduced fertility because the gametes (eggs or sperm) it produces will have too few or too many chromosomes, due to failure of meiosis. The hybrid may also be infertile for other, more complex reasons. Natural selection will then tend to favour isolating mechanisms which act earlier on in the reproductive process to prevent such hybrids being produced.*]

At one time it appeared to me probable that the sterility of first crosses and of hybrids might have been slowly acquired through the natural selection of slightly lessened degrees of fertility which, like any other variation, spontaneously appeared in certain individuals of one variety when crossed with those of another variety. For it would clearly be advantageous to two varieties or incipient species if they could be kept from blending. In the first place, it may be remarked that species inhabiting distinct regions are often sterile when crossed; now it could clearly have been of no advantage to such separated species to have been rendered mutually sterile, and consequently this could not have been effected through natural selection; but it may perhaps be argued that, if a species was rendered sterile with some one compatriot, sterility with other species would follow as a necessary contingency. In the second place, it is

almost as much opposed to the theory of natural selection as to that of special creation, that in reciprocal crosses the male element of one form should have been rendered utterly impotent on a second form, whilst at the same time the male element of this second form is enabled freely to fertilise the first form; for this peculiar state of the reproductive system could hardly have been advantageous to either species.

In considering the probability of natural selection having come into action in rendering species mutually sterile, the greatest difficulty will be found to lie in the existence of many graduated steps from slightly lessened fertility to absolute sterility. It may be admitted that it would profit an incipient species if it were rendered in some slight degree sterile when crossed with its parent form or with some other variety; for thus fewer bastardised offspring would be produced to commingle their blood with the new species in process of formation. But after mature reflection it seems to me that this could not have been effected through natural selection. Take the case of any two species which, when crossed, produce few and sterile offspring; now what is there which could favour the survival of those individuals which happened to be endowed in a slightly higher degree with mutual infertility, and which thus approached by one small step towards absolute sterility? Yet an advance of this kind, if the theory of natural selection be brought to bear, must have incessantly occurred with many species, for a multitude are mutually quite barren.

Reciprocal Dimorphism

This subject will be found to throw some light on hybridism. Several plants present two forms, which exist in about equal numbers and which differ in no respect except in their reproductive organs; one form having a long pistil with short stamens, the other a short pistil with long stamens; the two having differently sized pollen-grains. Now I have shown that in order to obtain full

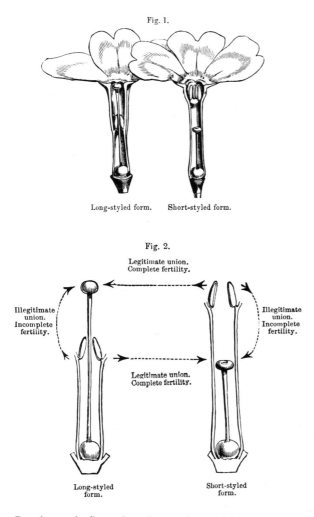

Fig. 1.

Long-styled form.   Short-styled form.

Fig. 2.

Legitimate union.
Complete fertility.

Illegitimate
union.
Incomplete
fertility.

Illegitimate
union.
Incomplete
fertility.

Legitimate union.
Complete fertility.

Long-styled
form.   Short-styled
form.

Darwin was the first to investigate and explain the phenomenon of dimorphism in flowers such as primroses and cowslips. Each plant always produces the same sort of flowers, but these can only be pollinated by flowers of the other type. Several mechanisms ensure this. Firstly, when an insect enters a long-styled flower the pollen is dusted onto the front of its body, whereas the pollen of a short-styled flower is dusted onto its rear end. When it visits other flowers, long-styled flower pollen will be transferred onto short styles and vice versa. Secondly, as Darwin discovered by painstaking experimentation, even if the pollen does fall onto the wrong sort of style, fertilization is difficult and the offspring are often sterile. The purpose of this, in an evolutionary sense, is to maintain outcrossing which is favourable to variability. Darwin devoted an entire book to this subject, *The Different Forms of Flowers on Plants of the Same Species*, published in 1877, from which these diagrams of the cowslip are taken. He wrote in his autobiography: *'No little discovery of mine ever gave me so much pleasure as the making out of the meaning of heterostyled flowers.'*

fertility with these plants, it is necessary that the stigma of the one form should be fertilised by pollen taken from the stamens of the other form. So two unions, which may be called legitimate, are fully fertile; and two, which may be called illegitimate, are more or less infertile. The infertility which may be observed when they are illegitimately fertilised differs much in degree, up to absolute and utter sterility, in the same manner as occurs in crossing distinct species.

In these respects, and in others which might be added, the forms of the same undoubted species when illegitimately united behave in exactly the same manner as do two distinct species when crossed. This led me carefully to observe, during four years, many seedlings raised from several illegitimate unions. The chief result is that these illegitimate plants, as they may be called, are not fully fertile.

It is possible to raise both long-styled and short-styled illegitimate plants. These can then be properly united in a legitimate manner. When this is done, there is no apparent reason why they should not yield as many seeds as did their parents when legitimately fertilised. But such is not the case. They are all infertile, in various degrees; some being so utterly and incurably sterile that they did not yield during four seasons a single seed or even seed-capsule. The sterility of these illegitimate plants, when united with each other in a legitimate manner, may be strictly compared with that of hybrids when crossed *inter se*. It is hardly an exaggeration to maintain that illegitimate plants are hybrids, produced within the limits of the same species by the improper union of certain forms.

These facts are important because they show us, first, that the physiological test of lessened fertility is no safe criterion of specific distinction; secondly that there is some unknown bond which connects the infertility of illegitimate unions with that of their illegitimate offspring, so that we are led to extend the same view to first crosses and hybrids.

We may infer as probable from the consideration of dimorphic plants, that the sterility of distinct species when crossed and of their hybrid progeny depends exclusively on the nature of their sexual elements, and not on any difference in their structure or general constitution.

Fertility of Varieties when Crossed, and of their Mongrel Offspring, not universal

It may be urged that there must be some essential distinction between species and varieties, inasmuch as the latter, however much they may differ from each other in external appearance, cross with perfect facility and yield perfectly fertile offspring. With some exceptions, presently to be given, I fully admit that this is the rule. But the subject is surrounded by difficulties, for, looking to varieties produced under nature, if two forms hitherto reputed to be varieties be found in any degree sterile together, they are at once ranked by most naturalists as species. For instance, the blue and red pimpernel, which are considered by most botanists as varieties, are said by Gärtner to be quite sterile when crossed, and he consequently ranks them as undoubted species.

If we turn to varieties produced under domestication we are still involved in some doubt. The perfect fertility of so many domestic races differing widely from each other in appearance, for instance those of the pigeon or the cabbage, is a remarkable fact; more especially when we reflect how many species there are which, though resembling each other most closely, are utterly sterile when intercrossed. However, it may be observed that the amount of external difference between two species is no sure guide to their degree of mutual sterility, so that similar differences in the case of varieties would be no sure guide. It is certain that with species the cause lies exclusively in differences in their sexual constitution.

The real difficulty in our present subject is not, as it appears to me, why domestic varieties have not become mutually infertile when crossed, but why this has so generally occurred with natural varieties as soon as they have been permanently modified in a sufficient degree to take rank as species. We are far from precisely knowing the cause; nor is this surprising, seeing how profoundly ignorant we are in regard to the normal and abnormal action of the reproductive system. But we can see that species will have been exposed during long periods of time to more uniform conditions than have domestic varieties; and this may well make a wide difference in the result.

I have as yet spoken as if the varieties of the same species were invariably fertile when intercrossed. But Gärtner kept during several years a dwarf kind of maize with yellow seeds and a tall variety with red seeds growing near each other in his garden; and although these plants have separated sexes, they never naturally crossed. He then fertilised thirteen flowers of the one kind with pollen of the other; but only a single head produced any seed, and this one head produced only five grains. The hybrid plants thus raised were themselves *perfectly* fertile; so that even Gärtner did not consider the two varieties as specifically distinct.

We may conclude that fertility does not constitute a fundamental distinction between varieties and species when crossed. The general sterility of crossed species may safely be looked at, not as a special acquirement or endowment, but as incidental on changes of an unknown nature in their sexual elements. Finally, then, although we are ignorant of the precise cause of the sterility of first crosses and of hybrids, yet the facts given in this chapter do not seem to me opposed to the belief that species aboriginally existed as varieties.

Chapter 10
On the Imperfection
of the Geological Record

In the sixth chapter I enumerated the chief objections against my views, including the distinctness of specific forms and their not being blended together by innumerable transitional links. The main cause of links not now occurring everywhere depends on the very process of natural selection, through which new varieties continually supplant their parent-forms. But the number of intermediate varieties which have formerly existed must be truly enormous. Why then is not every geological formation full of such intermediate links? Geology assuredly does not reveal any such finely graduated organic chain; and this, perhaps, is the most serious objection which can be urged against the theory. The explanation lies, as I believe, in the extreme imperfection of the geological record.

It should always be borne in mind what sort of intermediate forms formerly existed. I have found it difficult to avoid picturing forms *directly* intermediate between two species. But this is a wholly false view; we should always look for forms intermediate between each species and a common but unknown progenitor, generally different from all its modified descendants. Thus the fantail and pouter pigeons are both descended from the rock-pigeon; if we possessed all intermediate varieties, we should have an extremely close series between both and the rock-pigeon; but we should have no varieties directly intermediate between the fantail and pouter, none combining a tail somewhat expanded with a crop somewhat enlarged: the characteristic features of these two breeds. These two breeds, moreover, have become so modified that, if we

had no evidence regarding their origin, it would not be possible to determine whether they had descended from the rock-pigeon or from some allied form.

With natural species, such as the horse and tapir, the common parent will have had much resemblance to them, but in some points of structure may have differed considerably from both, even perhaps more than they differ from each other. Hence we should be unable to recognise the parent-form of any two species unless we had a nearly perfect chain of intermediate links.

It is just possible by the theory, that one of two living forms might have descended from the other; for instance a horse from a tapir; and in this case *direct* intermediate links will have existed between them. But such a case would imply that one form had remained for a very long period unaltered, whilst its descendants had undergone a vast amount of change; and the principle of competition between child and parent will render this a very rare event.

On the Lapse of Time, as inferred from the rate of Deposition and extent of Denudation
Independently of our not finding fossil remains of infinitely numerous links, it may be objected that time cannot have sufficed for so great an amount of organic change, all changes having been effected slowly. It is hardly possible for me to recall to the reader who is not a practical geologist the facts leading the mind feebly to comprehend the lapse of time. He who can read Sir Charles Lyell's grand work on the Principles of Geology, which the future historian will recognise

as having produced a revolution in natural science, and yet does not admit how vast have been the past periods of time, may at once close this volume. We can best gain some idea of past time by knowing the agencies at work: for the thickness of our sedimentary formations is the measure of the denudation which the earth's crust has elsewhere undergone. Therefore a man should examine the great piles of superimposed strata, and watch the rivulets bringing down mud, in order to comprehend something about the duration of past time.

It is good to wander along the coast and mark the process of degradation. The tides in most cases reach the cliffs only for a short time twice a day, and the waves eat into them only when they are charged with sand or pebbles. At last the base of the cliff is undermined, huge fragments fall down, and these have to be worn away atom by atom, until they can be rolled about by the waves, then more quickly ground into pebbles, sand, or mud. But how often do we see, along the bases of retreating cliffs, rounded boulders, all thickly clothed by marine productions, showing how little they are abraded and how seldom they are rolled about!

We have, however, recently learnt that much more important than the waves is the chemical action of rainwater with its dissolved carbonic acid, and in colder countries frost, which degrade the whole surface of the land. On a rainy day, even in a gently undulating country, we see the effects of this subaerial degradation in the muddy

'It is hardly possible for me to recall to the reader who is not a practical geologist the facts leading the mind feebly to comprehend the lapse of time . . . We can best gain some idea of past time by knowing the agencies at work: for the thickness of our sedimentary formations is the measure of the denudation which the earth's crust has elsewhere undergone. Therefore a man should examine the great piles of superimposed strata, and watch the rivulets bringing down mud, in order to comprehend something about the duration of past time.' **In the Grand Canyon the Colorado River has sliced downwards through rocks which represent over 200 million years of sedimentary deposits: the rocks at the top of the canyon date from the Permian, those at the bottom are Precambrian.**

rills which flow down every slope. The great lines of escarpment in the Wealden district owe their origin in part to the rocks of which they are composed having resisted subaerial denudation better than the surrounding surface; this surface consequently has been gradually lowered, with the lines of harder rock left projecting. Nothing impresses the mind with the vast duration of time more forcibly than the conviction that subaerial agencies, which seem to work so slowly, have produced great results.

When thus impressed with the slow rate at which the land is worn away through subaerial and coastal action, it is good, in order to appreciate the past duration of time, to consider the thickness of our sedimentary formations. I remember having been much struck when viewing volcanic islands which have been worn by the waves into perpendicular cliffs of one or two thousand feet in height; for the gentle slope of the lava-streams, due to their formerly liquid state, showed at a glance how far the hard, rocky beds had once extended into the open ocean.

In all parts of the world the piles of sedimentary strata are of wonderful thickness. In the Cordillera I estimated one mass of conglomerate at ten thousand feet; and although conglomerates have probably been accumulated at a quicker rate than finer sediments, yet from being formed of worn pebbles, each of which bears the stamp of time, they are good to show how slowly the mass must have been heaped together. Professor Ramsay has given me the maximum thickness of the successive formations in *different* parts of Great Britain:

Palaeozoic strata
(not including igneous beds)   57,154 feet
Secondary [Mesozoic] strata   13,190 feet
Tertiary strata   2,240 feet

making altogether 72,584 feet; that is, nearly thirteen and three-quarters miles. Between successive formations we have, in the opinion of

'In all parts of the world the piles of sedimentary strata are of wonderful thickness. In the Cordillera I estimated one mass of conglomerate at ten thousand feet . . .' The Andes consist of three chains of mountains, the Western Cordillera, the Central Cordillera and the Eastern Cordillera, although all three do not run the full length of the range. Special names are also given to certain sections of each of these mountain chains, such as the Cordillera Blanca in Peru. Darwin, however, uses 'Cordillera' in a general way, to refer to the Andes as a whole. Shown here is the valley between Nevado de Huascarán and Nevado de Huandoy in the Cordillera Blanca.

most geologists, blank periods of enormous length, so that this lofty pile gives but an inadequate idea of the time elapsed during its accumulation. The consideration of these facts impresses the mind almost in the same manner as does the vain endeavour to grapple with the idea of eternity.

Nevertheless this impression is partly false. Mr. Croll remarks that we do not err 'in forming too great a conception of the length of geological periods', but in estimating them by years. He shows, by calculating the known amount of sediment annually brought down by certain rivers, that 1000 feet of solid rock, as it became gradually disintegrated, would thus be removed from the mean level of the whole area in the course of six million years. This seems an astonishing result, and some considerations lead to the suspicion that it may be too large, but even if halved or quartered it is still very surprising. Few of us, however, know what a million really means. Mr. Croll gives the following illustration: take a strip of paper, 83 feet 4 inches in length, and stretch it along the wall of a large hall; then mark off at one end the tenth of an inch. This tenth of an inch will represent one hundred years, and the entire strip a million years.

But let it be borne in mind, in relation to the subject of this work, what a hundred years implies, represented as it is by a measure utterly insignificant in this hall. Several eminent breeders, during a single life-time, have so largely modified some of the higher animals, which propagate their kind much more slowly than most of the lower animals, that they have formed a new sub-breed; and a hundred years represents the work of two breeders. Species, however, change much more slowly, and within the same country only a few change at the same time. For new places in nature do not occur until after long intervals, due to physical changes or the immigration of new forms. Moreover variations of the right nature would not always occur at once. Unfortunately we have no means of determining how long it takes to modify a species.

On the Poorness of our Palaeontological Collections

Now let us turn to our richest geological museums, and what a paltry display we behold! That our collections are imperfect is admitted by everyone. Many fossil species are known from single and often broken specimens. Only a small portion of the earth has been geologically explored, and no part with sufficient care. Shells and bones decay and disappear when left on the bottom of the sea where sediment is not accumulating. We err

when we assume that sediment is being deposited over the whole bed of the sea sufficiently quickly to embed fossil remains. Throughout an enormously large proportion of the ocean the bright blue tint bespeaks its purity. The remains which do become embedded, if in sand or gravel, will, when the beds are upraised, generally be dissolved by rain-water charged with carbonic acid.

Some of the animals which live on the beach between high- and low-water mark seem to be rarely preserved. For instance, the several species of the Chthamalinae (a sub-family of sessile cirripedes) coat the rocks all over the world: they are all strictly littoral, with the exception of a single deep-water species, and a fossil of this species has been found in Sicily, whereas not one other species has been found in any Tertiary formation: yet it is known that the genus *Chthamalus* existed during the Chalk [Cretaceous] period.

Lastly, many great deposits, requiring a vast length of time for their accumulation, are entirely destitute of organic remains, without our being able to assign any reason: one of the most striking instances is that of the Flysch formation, which consists of shale and sandstone several thousand feet thick, extending from Vienna to Switzerland; throughout this great mass no fossils, except a few vegetable remains, have been found.

With respect to the terrestrial productions of the Secondary and Palaeozoic periods, it is superfluous to state that our evidence is fragmentary in an extreme degree. In regard to mammiferous remains, a glance at the historical table in Lyell's Manual will bring home the truth how accidental and rare is their preservation. Nor is their rarity surprising when we remember how large a proportion of the bones of Tertiary mammals have been discovered in caves or lacustrine deposits, and that not a cave or true lacustrine bed is known from our Secondary [Mesozoic] or Palaeozoic periods.

But the imperfection in the geological record largely results from another and more important

In 1938, a live coelacanth was discovered off the coast of Africa. This fish was known from fossils, but there was no trace of it in the fossil record since the Cretaceous, and it was presumed to have been extinct for a hundred million years. The fact that it had been alive throughout this period, but had left no fossils as far as is known, bears out Darwin's contention that the fossil record is very imperfect.

cause, namely from the several formations being separated from each other by wide intervals of time. When we see the formations it is difficult to avoid believing that they are closely consecutive. But we know, for instance, from Sir R. Murchison's great work on Russia, what wide gaps there are in that country between the superimposed formations; so it is in North America and in many other parts of the world. The most skilful geologist would never have suspected that, during the periods which were barren in his own country, great piles of sediment charged with new forms of life had elsewhere been accumulated. The frequent and great changes in the mineral composition of consecutive formations, implying

great changes in the surrounding lands whence the sediment was derived, accord with the belief of vast intervals of time having elapsed between each formation.

We can see why the geological formations of each region have not followed each other in close sequence. Scarcely any fact struck me more when examining many hundred miles of the South American coasts, which have been upraised several hundred feet within the Recent [Holocene, *but see Glossary*] period, than the absence of any Recent deposits sufficiently extensive to last for even a short geological period. Along the whole west coast Tertiary beds are so poorly developed that no record of several successive marine faunas will probably be preserved to a distant age. Along this rising coast no extensive formations with Recent or Tertiary remains can anywhere be found, though the supply of sediment must have been great from the enormous degradation of the coast-rocks and from muddy streams entering the

'We can see why the geological formations of each region have not followed each other in close sequence. Scarcely any fact struck me more when examining many hundred miles of the South American coasts, which have been upraised several hundred feet within the Recent period, than the absence of any Recent deposits sufficiently extensive to last for even a short geological period . . . The explanation is that as soon as they are brought up by the slow rising of the land the littoral and sublittoral deposits are continually worn away by the grinding action of the coast-waves.' **Part of the coast of Peru, showing the steady erosion of the land by wave action.**

sea. The explanation is that as soon as they are brought up by the slow rising of the land the littoral and sub-littoral deposits are continually worn away by the grinding action of the coast-waves.

Sediment must be accumulated in extremely thick or extensive masses to withstand the incessant action of the waves when first upraised, as well as the subsequent subaerial degradation. Such thick accumulations may be formed in two ways, either in profound depths of the sea, in which case the bottom will not be inhabited by so many forms of life as the more shallow seas, and

the mass when upraised will give an imperfect record of the organisms which existed in the neighbourhood. Or sediment may be deposited over a shallow bottom, if it continues slowly to subside; as long as the rate of subsidence and supply of sediment nearly balance each other, the sea will remain shallow and favourable for many forms, and thus a rich fossiliferous formation, thick enough to resist a large amount of denudation, may be formed. I am convinced that nearly all our ancient formations which are rich in fossils throughout their thickness, have thus been formed during subsidence.

Each area has undergone numerous slow oscillations of level, and these oscillations have affected wide spaces. Consequently, formations rich in fossils and sufficiently thick to resist subsequent degradation will have been formed over wide spaces during periods of subsidence, but only where the supply of sediment was sufficient to keep the sea shallow and to embed the remains before they decayed. On the other hand, as long as the bed of the sea remained stationary, *thick* deposits cannot have accumulated in the shallow parts, which are the most favourable to life; beds accumulated during elevation will generally have been destroyed by being brought within the limits of the coast-action.

These remarks apply chiefly to littoral and sublittoral deposits. In the case of a shallow sea, such as that within the Malay Archipelago, a formation might be formed during elevation and yet not suffer excessively from denudation during its slow upheaval; but the thickness could not be great, for it would be less than the depth in which it was formed; nor would the deposit be much consolidated, nor be capped by overlying formations, so that it would run a good chance of being worn away by atmospheric degradation and the sea during subsequent oscillations. However, if one area subsided after rising and before being denuded, the deposit formed during the rising, though not thick, might afterwards become protected by fresh accumulations, and thus be preserved.

Furthermore our present metamorphic schists and plutonic rocks have been stripped of their covering to an enormous extent. For such rocks could not have crystallised whilst uncovered. Admitting then that gneiss, mica-schist, granite, diorite, etc., were once covered up, how can we account for the naked and extensive areas of such rocks except on the belief that they have subsequently been completely denuded of all overlying strata? That such extensive areas exist throughout the world cannot be doubted. South of the Amazon is an area composed of rocks of this nature equal to Spain, France, Italy, part of Germany, and the British Islands, all conjoined. Turning to the United States and Canada, I find that the metamorphic and granitic rocks exceed the whole of the newer Palaeozoic formations. Thus it is probable that in some parts of the world whole formations have been completely denuded, with not a wreck left behind.

During periods of elevation the area of land and adjoining shallow parts of the sea will be increased – circumstances favourable for the formation of new varieties and species; but during such periods there will generally be a blank in the geological record. On the other hand during subsidence the inhabited area and number of inhabitants will decrease, and consequently few new varieties will be formed; and it is during these very periods of subsidence that the deposits which are richest in fossils have been accumulated.

On the Absence of Numerous Intermediate Varieties in any Single Formation

It cannot be doubted that the geological record, viewed as a whole, is extremely imperfect; but if we confine our attention to any one formation, it becomes much more difficult to understand why we do not find closely graduated varieties between the allied species which lived at its commencement and at its close. Although each formation has

indisputably required a vast number of years for its deposition, several reasons can be given why each should not commonly include a graduated series of links.

Probably each formation is short compared with the period requisite to change one species into another. When we see a species first appearing in the middle of any formation, it would be rash to infer that it had not elsewhere previously existed. So again when we find a species disappearing before the last layers have been deposited, it would be equally rash to suppose that it then became extinct.

When we see a species first appearing in any formation, the probability is that it only then first immigrated into that area. For instance, several species appeared earlier in the Palaeozoic beds of North America than in Europe, time having apparently been required for their migration from the American seas. In examining the latest deposits in various quarters, it has everywhere been noted that some still existing species are common in the deposit, but have become extinct in the immediately surrounding sea, or, conversely, that some are now abundant in the neighbouring sea, but are absent in this deposit. Reflect on the amount of migration of the inhabitants of Europe during the glacial epoch, which forms only a part of one geological period; and likewise reflect on the changes of level and climate, and on the great lapse of time, all included within this same epoch. Yet it may be doubted whether, in any quarter of the world, sedimentary deposits *including fossil remains* have accumulated within the same area during the whole of this period.

In order to get a perfect gradation between two forms in the upper and lower parts of the same formation, the deposit must have gone on continuously accumulating during a long period; hence the deposit must be very thick, and the species undergoing change must have lived in the same district throughout the whole time. But we have seen that a thick formation, fossiliferous throughout, can accumulate only during a period of subsidence; and to keep the depth approximately the same, which is necessary that the same species may live on the same space, the supply of sediment must nearly counterbalance the amount of subsidence. But this same movement of subsidence tends to submerge the area whence the sediment is derived, and thus diminish the supply. In fact, this nearly exact balancing is probably a rare contingency; for very thick deposits are

One difficulty Darwin faced over the geological record was the absence of evidence of gradual change in any species between the lower and upper layers of the same stratum. Careful search has since revealed some examples of this, as in the fossil heart-urchin genus *Micraster*. Specimens from different parts of the Cretaceous show a steady change in shape, in the position of the mouth and anus on the lower side *(left)* and the apical disc on the upper side *(right)*.

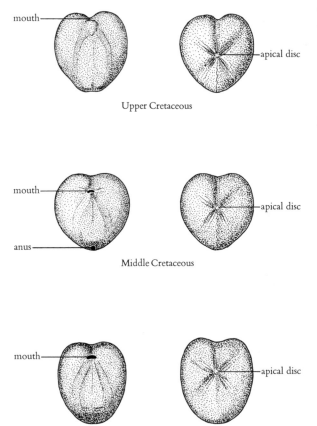

mouth

apical disc

Upper Cretaceous

mouth

apical disc

anus

Middle Cretaceous

mouth

apical disc

Lower Cretaceous

usually barren of organic remains, except near their limits.

It would seem that each separate formation has generally been intermittent in its accumulation. The common case of a formation composed of beds of widely different minerals suggests that the process of deposition has been interrupted. Nor will the closest inspection of a formation give us any idea of the length of time which its deposition may have consumed. Many beds only a few feet in thickness represent formations elsewhere thousands of feet thick, which must have required an enormous period for their accumulation. Hence, when the same species occurs at the bottom, middle and top of a formation, the probability is that it has not lived on the same spot during the whole period, but has disappeared and reappeared, perhaps many times, during the same geological period. Consequently if it were to undergo modification during this deposition, a section would not include all the fine gradations which must have existed, but abrupt changes of form.

Naturalists have no golden rule by which to distinguish species and varieties; they grant some little variability to each species, but when they meet with a somewhat greater amount of difference between any two forms, they rank both as species, unless they can connect them by the closest gradations. Thus in a formation we might obtain the parent-species and its modified descendants, and unless we obtained numerous transitional gradations we should not recognise their blood-relationship, and should consequently rank them as distinct species.

It is notorious on what slight differences many palaeontologists have founded their species. Look again at the later Tertiary deposits, which include many shells believed by the majority of naturalists to be identical with existing species; but some excellent naturalists, as Agassiz and Pictet, maintain that all these Tertiary species are specifically distinct, though the distinction is admitted to be very slight; so that here, unless we believe that these eminent naturalists have been misled by their imaginations, and that these Tertiary species really present no difference whatever from their living representatives, or unless we admit, in opposition to most naturalists, that these Tertiary species are all truly distinct from the Recent [Holocene, *but see Glossary*] we have evidence of slight modifications of the kind required.

It is a more important consideration that the period during which each species underwent modification, though long as measured by years, was probably short in comparison with that during which it remained unchanged. This would greatly lessen the chance of our being able to trace the stages of transition in any one geological formation.

It should not be forgotten that at the present day, with perfect specimens for examination, two forms can seldom be connected by intermediate varieties until many specimens are collected from many places; and with fossil species this can rarely be done. We shall, perhaps, best perceive the difficulty of connecting species by numerous, intermediate fossil links by asking ourselves whether future geologists will be able to prove that our different breeds of cattle, horses and dogs are descended from a single stock or from several aboriginal stocks. This could be effected only by the discovery in a fossil state of numerous gradations; and such success is improbable in the highest degree.

It has been asserted by writers who believe in the immutability of species that geology yields no linking forms. This assertion is certainly erroneous. If we take a genus having a score of species and destroy four-fifths of them, no one doubts that the remainder will stand much more distinct from each other. If the extreme forms in the genus happen to have been thus destroyed, the genus itself will stand more distinct from other allied genera. What geological research has not revealed is the former existence of infinitely numerous gradations connecting nearly all existing and extinct

species. But this ought not to be expected; yet this has been repeatedly advanced as a most serious objection against my views.

On the sudden Appearance of whole Groups of allied Species

The abrupt manner in which whole groups of species suddenly appear in certain formations has been urged as a fatal objection to the belief in the transmutation of species. If numerous species belonging to the same genera or families have really started into life at once, the fact would be fatal to the theory of evolution through natural selection. For the development of a group of forms, all descended from one progenitor, must have been an extremely slow process; and the progenitors must have lived long before their modified descendants. But we continually overrate the perfection of the geological record and falsely infer, because certain genera or families have not been found beneath a certain stage, that they did not exist before that stage. Such negative evidence is worthless, as experience has so often shown. We continually forget how large the world is compared with the area over which our geological formations have been carefully examined. We do not make due allowance for the intervals of time elapsed between consecutive formations. These intervals will have given time for the multiplication of species from some one parent-form; and in the succeeding formation such species will appear as if suddenly created.

I will give a few examples to show how liable we are to error in supposing that groups of species have suddenly been produced. Even in so short an interval as that between the first and second editions of Pictet's great work on Palaeontology, published in 1844–46 and in 1853–57, the conclusions on the first appearance and disappearance of several groups of animals have been considerably modified. In geological treatises published not many years ago, mammals were always spoken of as having abruptly come in at the commencement of the Tertiary series. And now one of the richest known accumulations of fossil mammals belongs to the middle of the Secondary [Mesozoic] series; and true mammals have been discovered in the new red sandstone at nearly the commencement of this great series. Cuvier used to urge that no monkey occurred in any Tertiary stratum; but now extinct species have been discovered in India, South America and Europe as far back as the Miocene stage. Had it not been for the rare accident of the preservation of footsteps in the new red sandstone of the United States, who would have supposed that at least thirty different bird-like animals, some of gigantic size, existed during that period? [*These were in fact the footprints of dinosaurs.*] Not long ago, palaeontologists maintained that the whole class of birds came suddenly into existence during the Eocene period; but now we know that a bird certainly lived during the deposition of the upper greensand [Cretaceous period]; and still more recently that strange bird the Archaeopteryx, with a long, lizard-like tail, bearing a pair of feathers on each joint, and with its wings furnished with two free claws, has been discovered. Hardly any recent discovery shows more forcibly than this how little we as yet know of the former inhabitants of the world.

The case most frequently insisted on by palaeontologists of the apparently sudden appearance of a whole group of species is that of the teleostean fishes low down in the Chalk [Cretaceous] period. But certain Jurassic and Triassic forms are now commonly admitted to be teleostean; and even some Palaeozoic forms have thus been classed. If the teleosteans had really appeared suddenly in the northern hemisphere at the commencement of the chalk formation, the fact would have been highly remarkable; but it would not have formed an insuperable difficulty, unless it could likewise have been shown that at the same period the species were suddenly developed in other quarters of the world. Hardly any fossil-fish are known from south of the equator, and very few species

'*Had it not been for the rare accident of the preservation of footsteps in the new red sandstone of the United States, who would have ventured to suppose that at least thirty different bird-like animals, some of gigantic size, existed during that period?*' In fact these footprints, which the religious palaeontologists of the nineteenth century attributed to a creature they called 'Noah's Raven', are now known to be those of dinosaurs which ran on their hindlegs alone and had bird-like feet.

are known from formations in Europe. Some few families of fish now have a confined range; the teleostean fishes might formerly have had a similarly confined range, and after having been largely developed in some one sea, have spread widely.

On the sudden Appearance of Groups of allied Species in the lowest known Fossiliferous Strata

Another and allied difficulty, which is much more serious, is the sudden appearance of species belonging to several of the main divisions of the animal kingdom in the lowest known fossiliferous rocks, namely those of the Cambrian and Silurian formations. It cannot be doubted that all the Cambrian and Silurian trilobites are descended from some one crustacean, which must have lived long before the Cambrian age, and which probably differed greatly from any known animal. If our theory be true, it is indisputable that before the lowest Cambrian stratum was deposited long periods elapsed, and that during these periods the world swarmed with living creatures. Here we encounter a formidable objection; for it seems doubtful whether the

earth, in a fit state for the habitation of living creatures, has lasted long enough. Sir W. Thomson concludes that the consolidation of the crust can hardly have occurred less than 20 or more than 400 million years ago, but probably not less than 98 or more than 200 million years. These very wide limits show how doubtful the data are. Mr. Croll estimates that about 60 million years have elapsed since the Cambrian period, but this, judging from the small amount of organic change since the commencement of the Glacial epoch, appears a very short time for the mutations of life which have certainly occurred since the Cambrian formation; and the previous 140 million years can hardly be considered as sufficient for the development of the varied forms of life which already existed during the Cambrian period. It is, however, probable that the world at a very early period was subjected to more rapid and violent changes than those now occurring, and such changes would have induced changes at a corresponding rate in the organisms which then existed.

To the question why we do not find rich fossiliferous deposits belonging to these assumed earliest periods prior to the Cambrian system, I can give no satisfactory answer. Several eminent geologists,

(left) Trilobites were hard-shelled animals, related to the crustaceans (crabs and shrimps), which lived on the sea bottom, and became extinct in the late Permian. They were abundant during the Cambrian and there were also many different types, which Darwin considered must have started to evolve from a common ancestor long before the beginning of the Cambrian. But extensive searches by palaeontologists have failed to reveal the Precambrian strata rich in fossils of multicellular animals which Darwin believed must somewhere exist. The problem of the so-called 'Cambrian explosion' is discussed in the introduction, on pp. 35–6.

(right) A rock from a Silurian deposit, containing numerous fossils of marine animals. Until the 1840s geologists believed that they had discovered the earliest creatures in the Silurian. The fact that new discoveries had pushed back the dawn of life from the Silurian to the Cambrian encouraged Darwin to believe that further geological exploration would extend the span of life on earth even further, as indeed it has.

with Sir R. Murchison at their head, were until recently convinced that we beheld in the organic remains of the lowest Silurian stratum the first dawn of life. But not very long ago M. Barrande added a lower stage, abounding with new species, beneath the Silurian system; and now, still lower down in the Lower Cambrian formation, Mr. Hicks has found in South Wales beds rich in trilobites and containing various molluscs and annelids.

[*The fossil which Darwin calls the Eozoon was later shown to be a crystalline formation, but careful search has more recently revealed numerous fossils of very simple, single-celled organisms in very ancient Precambrian rocks. However, the apparently rapid transition from these simple forms to the teeming, varied, and complex fauna of the Cambrian seas is still a matter of debate amongst palaeontologists. Darwin's hypothesis to explain the absence of fossiliferous Precambrian rocks does not fit in with current knowledge about continental drift (see maps on pp. 188–9).*]

Beneath the Silurian system in Canada are three great series of strata in the lowest of which the Eozoon is found. The Eozoon belongs to the most lowly organised of all classes of animals, but is highly organised for its class; it existed in countless numbers and certainly preyed on other minute organic beings, which must have lived in great numbers. Thus the words which I wrote in 1859 about the existence of living beings long before the Cambrian period have proved true.

Nevertheless, the difficulty of assigning any good reason for the absence of vast piles of strata

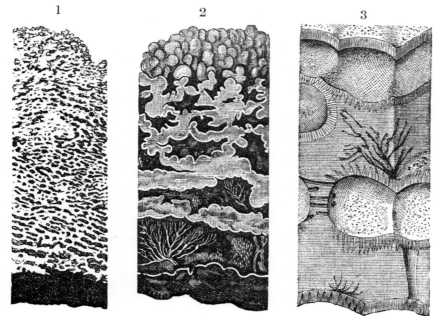

The discovery of numerous fossils of an animal named the Eozoon in Precambrian rocks in Canada delighted Darwin. Sadly, these 'fossils' were later shown to be nothing more than crystalline formations.

1. Portion of the serpentinous marble of Canada, composed of Eozoon; of the natural size. The broken black lines represent the serpentine, and the white spaces are the calcareous skeleton. Copied from Dr. Dawson's nature-printed section of a specimen of Eozoon Canadense (from Petite Nation Seigniory), first polished and then corroded with acid. [See Canadian Naturalist and Geologist, April, 1865.]

2. The serpentinous portion of a piece of Eozoon, after maceration in acid, magnified. It presents the natural casts of the chambers and tubes, or a model of the sarcode of the Eozoon. (After Dr. Carpenter's plate in the Intellectual Observer, No. XL. p. 300 &c.)

3. A portion of the chamber-walls, or calcareous shell, of the Eozoon, restored and highly magnified, showing the tubuliferous walls, the pseudopodial tufts in the intermediate skeleton, and the stolon-passages. (After Carpenter, *loc. cit.*)

rich in fossils beneath the Cambrian system is very great, and may be truly urged as a valid argument against the views here entertained. To show that it may hereafter receive some explanation, I will give the following hypothesis. The great oceans are mainly areas of subsidence, the great archipelagos areas of oscillations, and the continents areas of elevation. But we have no reason to assume that things have thus remained from the beginning of the world. At a period long antecedent to the Cambrian epoch, continents may have existed where oceans are now spread out; and oceans may have existed where our continents now stand. Nor should we be justified in assuming that if, for instance, the bed of the Pacific Ocean were now converted into a continent, we should there find sedimentary formations in a recognisable condition older than the Cambrian strata, supposing such to have been formerly deposited; for it might well happen that strata some miles nearer to the centre of the earth, which had been pressed on by an enormous weight of water, might have undergone far more metamorphic action than strata nearer to the surface.

The immense areas, for instance in South America, of naked metamorphic rocks, which must have been heated under great pressure, have always seemed to me to require some special explanation; perhaps we see in these large areas the many formations long anterior to the Cambrian epoch in a completely metamorphosed and denuded condition.

The several difficulties here discussed are all of the most serious nature. We see this in the fact that the most eminent palaeontologists, namely, Cuvier, Agassiz, Barrande, Pictet, Falconer, E. Forbes, etc., and all our greatest geologists, as Lyell, Murchison, Sedgwick, etc., have unanimously, often vehemently, maintained the immutability of species. But Sir Charles Lyell now gives the support of his high authority to the opposite side; and most geologists and palaeontologists are much shaken in their former belief.

Those who believe that the geological record is in any degree perfect will undoubtedly at once reject the theory. For my part, following out Lyell's metaphor, I look at the geological record as a history of the world imperfectly kept, and written in a changing dialect; of this history we possess the last volume alone, relating only to two or three countries. Of this volume, only here and there a short chapter has been preserved; and of each page, only here and there a few lines. Each word of the slowly changing language, more or less different in the successive chapters, may represent the forms of life which are entombed in our consecutive formations, and which falsely appear to us to have been abruptly introduced. On this view, the difficulties above discussed are greatly diminished or even disappear.

Chapter 11

On the Geological Succession of Organic Beings

Let us now see whether the several facts and laws relating to the geological succession of organic beings accord best with the common view of the immutability of species, or with that of their slow and gradual modification through variation and natural selection.

Every year tends to fill up the blanks between the Tertiary stages, and to make the proportion between the lost and existing forms more gradual. In some of the most recent beds only one or two species are extinct, and only one or two are new. The Secondary [Mesozoic] formations are more broken; but neither the appearance nor disappearance of the many species embedded in each formation has been simultaneous.

Species have not changed at the same rate or in the same degree. In the older Tertiary beds a few living shells may still be found in the midst of extinct forms. An existing crocodile is associated with many lost mammals and reptiles in the sub-Himalayan deposits. The Silurian *Lingula* differs but little from the living species of this genus, whereas most of the other Silurian molluscs and all the crustaceans have changed greatly. The productions of the land seem to have changed at a quicker rate than those of the sea. The amount of organic change is not the same in each successive formation. Yet if we compare any but the most closely related formations, all the species will be found to have undergone some change. When a species has once disappeared from the face of the earth, we have no reason to believe that the same identical form ever reappears.

These several facts accord well with our theory, which includes no fixed law of development causing all inhabitants of an area to change abruptly, or simultaneously, or to an equal degree. The process of modification must be slow, and will generally affect only a few species at the same time; for the variability of each species is independent of that of others. Whether variations which arise will be accumulated through natural selection, thus causing permanent modification, will depend on many complex contingencies – on variations being of a beneficial nature, on the freedom of intercrossing, on the slowly changing physical conditions of the country, on the immigration of new colonists, and on the nature of the other inhabitants with which the varying species come into competition. Hence one species can retain the same identical form much longer than others; or, if changing, can change in a less degree.

We find similar relations between the existing inhabitants of distinct countries; for instance, the land-shells and coleopterous insects of Madeira have come to differ considerably from those of Europe, whereas the marine shells and birds have remained unaltered. We can perhaps understand the apparently quicker rate of change in terrestrial productions compared with marine productions by their more complex organic and inorganic conditions of life, as explained in a former chapter. When many inhabitants have become improved, we can understand on the principle of competition that any form which did not become improved would be liable to extermination. Hence we see why all the species in the same region do at last become modified, for otherwise they would become extinct.

In members of the same class the average amount

of change during long periods of time may, perhaps, be nearly the same; but as our formations have been accumulated at wide and irregularly intermittent intervals, the amount of organic change exhibited by the fossils embedded in consecutive formations is not equal. Each formation, on this view, does not mark a new and complete act of creation, but only an occasional scene, taken almost at hazard, in an ever slowly changing drama.

We can clearly understand why a species when once lost should never reappear, even if the very same conditions of life should recur. For though the offspring of one species might be adapted to fill the place of another species in the economy of nature, yet the two forms would not be identically the same; for both would inherit different characters from their distinct progenitors; and organisms already differing would vary in a different manner. For instance, if all our fantail-pigeons were destroyed, fanciers might make a new breed hardly distinguishable from the present breed; but if the parent rock-pigeon were likewise destroyed, (and under nature parent-forms are generally supplanted by their improved offspring), a fantail identical with the existing breed could not be raised from any other species of pigeon or race of the domestic pigeon, for the successive variations would be different, and the newly formed variety would probably inherit from its progenitor some characteristic differences.

Groups of species follow the same general rules in their appearance and disappearance as do single species, changing more or less quickly and in a greater or lesser degree. A group, when it has once disappeared, never reappears. For all the species of the same group are the modified descendants of a common progenitor. As a rule each group shows a gradual increase in number until it reaches its maximum, and then, sooner or later, a gradual decrease. This gradual increase in number of the species of a group is strictly conformable with our theory, for the species and genera can increase

only slowly and progressively, the process of modification and the production of allied forms necessarily being a slow and gradual process – one species first giving rise to two or three varieties, these being slowly converted into species, which in their turn produce other varieties and species, and so on, like the branching of a great tree from a single stem, till the group becomes large.

On Extinction

On the theory of natural selection, the extinction of old forms and the production of new and improved forms are intimately connected together. The old notion of all the inhabitants of the earth having been swept away by catastrophes at successive periods is very generally given up. On the contrary, the Tertiary formations indicate that species and groups gradually disappear, one after another, first from one spot, then from another, and finally from the world. In some few cases, however, as by the final subsidence of an island, the process of extinction may have been rapid. No fixed law seems to determine the time during which any single species or genus endures. There is reason to believe that the extinction of a whole group is generally a slower process than its production: in some cases, however, the extermination of whole groups, as of ammonites towards the close of the Secondary [Mesozoic] period, has been wonderfully sudden.

The extinction of species has been involved in the most gratuitous mystery. Some authors have even supposed that, as the individual has a definite length of life, so have species a definite duration. No one can have marvelled more than I have done at the extinction of species. When I found in La Plata the tooth of a horse embedded with the remains of Mastodon, Megatherium, Toxodon, and other extinct monsters, which all co-existed

(overleaf) **While in South America Darwin unearthed fossils of strange extinct creatures, and noted that many of the living animals were unique to that continent. Subsequent research has pieced together a coherent picture of their evolution.**

1. South America became cut off from North America following the Palaeocene. The only representatives of placental mammals which were then present were archaic ungulates, which evolved to produce a large number of unique animals, including toxodon and macrauchenia. All these are now extinct. Primitive edentates were present which gave rise to the armadillos, anteaters and sloths. Although representatives of these still survive, many of their extinct relatives were much larger in size: the giant sloth megatherium was the size of an elephant. One group of edentates which are now extinct were the glyptodonts, a heavily-armoured offshoot of the armadillos. Marsupial mammals were then present in South America and since there were no placental carnivores, marsupial carnivores such as borhyaena evolved to fill this ecological niche. However the only surviving marsupials in the Americas are the small, shrew-like opossums.

2. During the Oligocene a string of islands allowed some small animals to migrate into South America from the North. The first set of invaders were Hystricomorph (porcupine suborder) rodents. From these evolved several rodents unique to South America, such as the capybara, the coypu, the bizcacha, the agouti and the South American porcupines. A later invasion of ancestral primates gave rise to the South American monkeys, which evolved independently of the New World monkeys: an instance of parallel evolution. Lastly came members of the racoon family, from which the coatis evolved. Meanwhile in North America the edentates, the marsupial mammals and the archaic ungulates became extinct in the face of competition from advanced placental mammals.

5. In the late Pliocene a complete land bridge began to emerge and major migrations of large animals took place. From North America came more advanced herbivores such as the horse, the llama, the peccary, mammoths and the mastodon. Competition for food resources from these advanced types probably contributed to the extinction of the archaic ungulates which had all disappeared by the close of the Pleistocene. Placental carnivores arriving from the north, such as the fox, the wolf, the sabre-tooth tiger and the jaguar may have been partly responsible for the extinction of these archaic ungulates, and it was almost certainly they who drove the marsupial carnivores, such as borhyaena to extinction. A less dramatic, but equally important invasion was made by shrews, rabbits, squirrels, and Cricetid (hamster family) rodents, the latter proving especially successful. Some South American mammals migrated into North America but none were successful for very long. Of these only armadillos, porcupines and opossums have survived to the present day. At the close of the Pleistocene many large mammals became extinct, not only in the Americas but throughout the world. The reason for this is not fully understood: rapid climatic changes associated with the Ice Ages may be implicated, but hunting by early man could have been the cause of these extinctions. The horse, mammoth, mastodon, sabre-tooth tigers and many others disappeared from both North and South America. The megatherium, which had survived the invasion of North American animals, disappeared at this time. The tapir, and certain other species survived in South America although they became extinct in North America.

Opossum

Megatherium

Toxodon

Macrauchenia

Borhyaena

Glyptodon

2

Porcupine

Agouti

Capybara

Coypu

Coati

Spider monkey

3

Mastodon

Jaguar

Horse

Llama

Wolf

Tapir

Cricetid rodent

Peccary

'When I found in La Plata the tooth of a horse embedded with the
remains of Mastodon, Megatherium, Toxodon, and other extinct
monsters . . . I was filled with astonishment; for seeing that the
horse, since its introduction by the Spaniards into South America,
has run wild over the whole country and has increased in numbers at
an unparalleled rate, I asked myself what could so recently have
exterminated the former horse under conditions of life apparently so
favourable. But my astonishment was groundless. Professor Owen
soon perceived that the tooth, though so like that of the existing
horse, belonged to an extinct species.' The disappearance of the
native horse species (which was very similar to the modern
horse), from South America cannot be as easily explained away
as Darwin suggests. Many large mammals died out in various
parts of the world at the close of the Pleistocene, for reasons
that are still not understood (see p. 168), and the extinction of
the horse in both North and South America appears to have
been part of this trend.

with still living shells at a very late geological
period, I was filled with astonishment; for seeing
that the horse, since its introduction by the
Spaniards into South America, has run wild over
the whole country and has increased in numbers
at an unparalleled rate, I asked myself what could
so recently have exterminated the former horse
under conditions of life apparently so favourable.
But my astonishment was groundless. Professor
Owen soon perceived that the tooth, though so

like that of the existing horse, belonged to an
extinct species. Had this horse been still living but
in some degree rare, we might have felt certain,
from the history of the naturalisation of the domestic horse in South America, that under more
favourable conditions it would in a very few
years have stocked the whole continent. But we
could not have told what the unfavourable conditions were which checked its increase, whether
some one or several contingencies. If the conditions had gone on becoming less and less favourable, the fossil horse would certainly have become
rarer and rarer, and finally extinct, its place being
seized on by some more successful competitor.

It is most difficult always to remember that the
increase of every creature is constantly checked
by unperceived hostile agencies, and that these
agencies are amply sufficient to cause rarity and
extinction. So little is this subject understood that
I have heard surprise repeatedly expressed at such
great monsters as the Mastodon and the Dinosaurs
having become extinct, as if mere bodily strength
gave victory in the battle of life. Mere size, on the

contrary, would in some cases determine quicker extermination from the greater amount of requisite food. Before man inhabited India or Africa, some cause must have checked the continued increase of the existing elephant. Dr. Falconer believes that it is chiefly insects which, from incessantly harassing and weakening the elephant, check its increase. It is certain that insects and bloodsucking bats determine the existence of the larger naturalised quadrupeds in several parts of South America.

We see in many cases in the more recent Tertiary formations that rarity precedes extinction; and we know that this has been the progress of events with those animals which have been exterminated through man's agency. Species generally become rare before they become extinct – to feel no surprise at the rarity of a species, and yet to marvel greatly when the species ceases to exist, is much the same as to feel no surprise at sickness, but, when the sick man dies, to wonder and to suspect that he died by some deed of violence.

The theory of natural selection is grounded on the belief that each new variety is produced by having some advantage over those with which it comes into competition; and the consequent extinction of the less-favoured forms almost inevitably follows. With our domestic productions, when an improved variety has been raised, it at first supplants the less improved varieties in the same neighbourhood; when much improved it is transported far and near, like our short-horn cattle, and takes the place of other breeds in other countries. Thus the appearance of new forms and the disappearance of old are bound together.

The competition will generally be most severe, as formerly explained, between forms which are most like each other in all respects. Hence improved descendants will generally cause the extermination of the parent-species and the nearest allies of that species. Thus a number of new species descended from one species, that is a new genus, comes to supplant an old genus belonging to the same family. But it must often have happened that a new species has seized the place occupied by a species belonging to a distinct group. If many allied forms be developed from the successful intruder, many will have to yield their places; and it will generally be the allied forms that will suffer from some inherited inferiority in common. In any case, a few of the sufferers may often be preserved for a long time, from being fitted to some peculiar station in life. Therefore the utter extinction of a group is generally a slower process than its production.

With respect to the apparently sudden extermination of whole families or orders, as of trilobites at the close of the Palaeozoic period and of ammonites at the close of the Secondary [Mesozoic] period, we must remember the probable wide intervals of time between our consecutive formations: in these intervals there may have been much slow extermination. Moreover when, by sudden immigration or by unusually rapid development, many species of a new group have taken possession of an area, many of the older species will have been exterminated in a correspondingly rapid manner.

Thus, as it seems to me, the manner in which species and groups become extinct accords well with the theory of natural selection. We need not marvel at extinction; if we must marvel, let it be at our own presumption in imagining that we understand the many complex contingencies on which the existence of each species depends. When we can precisely say why this species is more abundant than that, why this species and not another can be naturalised in a given country, then, and not until then, we may justly feel surprise why we cannot account for the extinction of any particular species or group of species.

On the Forms of Life changing almost simultaneously throughout the World

[*In recent years the theory of continental drift has been widely accepted. This theory states that the present continents once formed one large landmass and that they are still slowly drifting apart (see maps on pp. 188–9). This theory helps us to further understand the distribution of similar fossils in very old strata, but Darwin's discussion of the subject remains valid. The case of the Chalk or Cretaceous strata is an exceptional one, however, since both the rocks and the fossils within them are unusually similar in strata throughout the world for reasons which are not well understood.]*

Scarcely any palaeontological discovery is more striking than the fact that the forms of life change almost simultaneously throughout the world. Thus our European Chalk [Cretaceous] formation can be recognised in many distant regions, where not a fragment of the mineral chalk itself can be found, namely, in North and South America, at the Cape of Good Hope, and in India. For at these points the organic remains in certain beds present an unmistakable resemblance to those of the Chalk. It is not that the same species are met with; but they belong to the same families, genera and sections of genera. Moreover, forms which are not found in the Chalk of Europe but which occur in the formations either above or below, occur in the same order at these distant points of the world. In the several successive Palaeozoic formations of Russia, Western Europe and North America, a similar parallelism has been observed.

These observations, however, relate to marine inhabitants: we have not sufficient data to judge whether productions of the land and fresh water change in the same parallel manner. We may doubt whether they have thus changed: if the Megatherium, Macrauchenia and Toxodon had been brought to Europe from La Plata, without any information in regard to their geological position, no one would have suspected that they had co-existed with sea-shells all still living; but as these anomalous monsters co-existed with the Mastodon and Horse, it might at least have been inferred that they had lived during one of the later Tertiary stages.

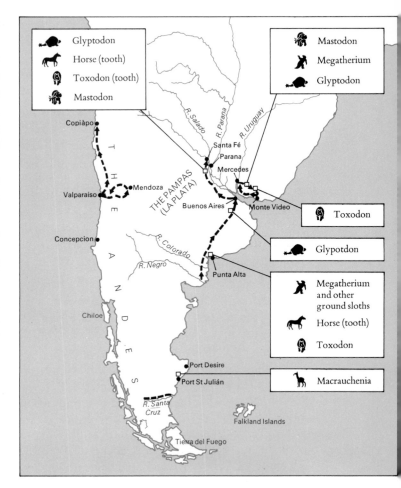

Whenever H.M.S. *Beagle* dropped anchor Darwin would explore the coastal region for fossils. Sometimes a prolonged stay in harbour allowed him to make a longer trip inland, or he would leave the *Beagle* at one port, travel overland, and meet up with the ship at another port. During some such inland excursions Darwin collected many particularly important fossils of extinct mammals. The map shows the excursions he made and the sites where fossils of extinct mammals were found.

When the marine forms of life are spoken of as having changed simultaneously, it must not be supposed that this expression relates to the same century, or even that it has a very strict geological sense. Nevertheless, all the more modern formations, namely the upper Pliocene, the Pleistocene and strictly modern beds of Europe, North and South America, and Australia, contain fossil remains in some degree allied, and do not include forms which are found only in the older underlying deposits.

This great fact of the parallel succession of the forms of life throughout the world is explicable on the theory of natural selection. New species are formed by having some advantage over older forms, and the forms which are already dominant give birth to the greatest number of new, dominant, and far-spreading varieties or incipient species. It is natural that they should spread and give rise in new countries to other new varieties. The process of diffusion would often be very slow, but in the course of time the dominant forms would generally spread and prevail. The diffusion would be slower with the terrestrial inhabitants of distinct continents than with the marine inhabitants of the continuous sea. We therefore find a less strict degree of parallelism in the succession of the productions of the land.

On the Affinities of Extinct Species to each other, and to Living Forms

Let us now look to the mutual affinities of extinct and living species. All fall into a few grand classes; and this fact is explained on the principle of descent. The more ancient any form is, the more, as a general rule, it differs from living forms. But extinct species can all be classed either in existing groups or between them, helping to fill up the intervals. As this statement has often been ignored or even denied it may be well to give some instances.

M. Gaudry has shown that many of the fossil mammals discovered by him in Attica serve to break down the intervals between existing genera. Cuvier ranked the Ruminants and Pachyderms as two of the most distinct orders of mammals; but so many fossil links have been disentombed that Owen has had to alter the whole classification, and has placed certain pachyderms in the same sub-order with ruminants; for example he dissolves by gradations the apparently wide interval between the pig and the camel. No one will deny

that Hipparion is intermediate between the existing horse and certain older ungulate forms. The Sirenia form a very distinct group of mammals, and one of the most remarkable peculiarities in the existing dugong and lamentin [manatee] is the entire absence of hind limbs; but the extinct Halitherium had an ossified thigh-bone, and it thus makes some approach to hoofed quadrupeds, to which the Sirenia are in other respects allied. The whales are widely different from all other mammals, but the Tertiary Zeuglodon and Squalodon are considered by Professor Huxley to be undoubtedly cetaceans, 'and to constitute connecting links with the aquatic carnivora.' Even the wide interval between birds and reptiles has been shown to be partially bridged over in the most unexpected manner by the extinct Archeopteryx and one of the Dinosaurs.

On the theory of descent with modification the main facts with respect to the mutual affinities of the extinct forms of life to each other and to living forms are explained in a satisfactory manner. And they are wholly inexplicable on any other view. On this same theory, it is evident that the fauna during any one great period in the earth's history will be intermediate in general character between that which preceded and that which succeeded it. I need give only one instance, namely the manner in which the fossils of the Devonian system, when first discovered, were at once recognised by palaeontologists as intermediate in character between those of the overlying Carboniferous and underlying Silurian systems. But each fauna is not necessarily exactly intermediate, as unequal intervals of time have elapsed between consecutive formations.

Closely connected with this statement is the fact that fossils from two consecutive formations are far more closely related to each other than are the fossils from two remote formations. On the theory of descent, the full meaning of the fossil remains from closely consecutive formations being closely related, though ranked as distinct species, is obvious. As the accumulation of each formation has often been interrupted, we ought to find closely allied forms after intervals very long as measured by years but only moderately long geologically; and these assuredly we do find. We find, in short, such evidence of the slow and scarcely sensible mutations of specific forms as we have the right to expect.

On the State of Development of Ancient compared with Living Forms

We have seen in the fourth chapter that the degree of differentiation and specialisation of the parts in organic beings is the best standard of their degree of perfection or highness, and natural selection will tend to render the organisation of each being more specialised; though leaving many creatures with simple structures fitted for simple conditions of life, and in some cases even degrading the organisation, yet leaving such degraded beings better fitted for new walks of life. Thus modern forms ought, on the theory of natural selection, to stand higher than ancient forms. Is this the case? It seems that this answer must be admitted as true, though difficult of proof.

It is no valid objection that certain Brachiopods have been but slightly modified from an extremely remote geological epoch, nor that Foraminifera have not progressed in organisation since the Laurentian epoch; for some organisms would have to remain fitted for simple conditions, and what could be better fitted for this end than these lowly organised Protozoa? When advanced up to any given point, there is no necessity on the theory of natural selection for their further continued progress, though they will, during each successive age, have to be slightly modified so as to hold their places in relation to slight changes in their conditions.

The problem whether organisation on the whole has advanced is in may ways excessively intricate. Even at the present day, looking to the same class, naturalists are not unanimous which

forms ought to be ranked as highest. To attempt to compare members of distinct types in the scale of highness seems hopeless: who will decide whether a cuttle-fish be higher than a bee – that insect which the great Von Baer believed to be 'in fact more highly organised than a fish, although upon another type'?

We ought also to compare the relative numbers at any two periods of the high and low classes throughout the world: if, for instance, at the present day fifty thousand kinds of vertebrate animals exist, and if we knew that at some former period only ten thousand kinds existed, we ought to look at this increase, which implies a great displacement of lower forms, as a decided advance in the organisation of the world.

On the Succession of the same Types within the same Areas, during the later Tertiary periods

Mr. Clift many years ago showed that the fossil mammals from the Australian caves were closely allied to the living marsupials of that continent. In South America a similar relationship is manifest in the gigantic pieces of armour, like those of the armadillo, found in several parts of La Plata; and Professor Owen has shown in the most striking manner that most of the fossil mammals buried there in such numbers are related to South American types. Professor Owen has extended this generalisation to the mammals of the Old World, and we see the same law in the extinct and gigantic birds of New Zealand.

Now what does this remarkable law of the succession of the same types within the same areas mean? On the theory of descent with modification it is at once explained; for the inhabitants of each quarter of the world will obviously tend to leave in that quarter closely allied though modified descendants. If the inhabitants of one continent formerly differed greatly from those of another continent, so will their modified descendants. But after very long intervals of time and great geographical changes, the feebler will yield to the more dominant, and there will be nothing immutable in the distribution of organic beings.

It may be asked in ridicule whether I suppose that the megatherium and other allied huge monsters, which formerly lived in South America, have left behind the sloth, armadillo, and anteater as their degenerate descendants. This cannot for an instant be admitted. These huge animals have become wholly extinct, and have left no progeny. But in the caves of Brazil there are many extinct species closely allied to the species still living in South America; and some of these fossils may have been the actual progenitors of the living species.

Fossilized remains of the glyptodon, an animal related to the present-day armadillos. Darwin writes: *'In South America a similar relationship is manifest in the gigantic pieces of armour, like those of the armadillo, found in several parts of La Plata; and Professor Owen has shown in the most striking manner that most of the fossil mammals buried there in such numbers are related to South American types.'* It was ironic that the comparative anatomist Richard Owen, who contributed so much to Darwin's ideas by his expert anatomical work on the fossils Darwin found in South America, later became his most vehement opponent among British scientists.

Summary of the preceding and present Chapters

I have attempted to show that the geological record is extremely imperfect; that only a small proportion of the globe has been geologically explored; that the number of specimens preserved is as nothing with the number which must have passed away; that owing to subsidence being almost necessary for the accumulation of deposits rich in fossils and thick enough to outlast future degradation, great intervals of time must have elapsed between successive formations, and there has probably been more extinction during periods of subsidence, and more variation during periods of elevation, when the record will have been least perfectly kept; that each single formation has not been continuously deposited, that the duration of each formation is probably short compared with the average duration of specific forms; that migration has played an important part in the first appearance of new forms in any one area; that varieties must have at first been local; and lastly, although each species must have passed through numerous transitional stages, it is probable that the

periods during which each underwent modification, though long as measured by years, have been short in comparison with the periods during which each remained unchanged. These causes will to a large extent explain why, though we do find many links, we do not find interminable varieties connecting together all extinct and existing forms by the finest graduated steps. It should also be borne in mind that any linking variety between two forms which might be found would be ranked, unless the whole chain could be perfectly restored, as a new and distinct species.

He who rejects this view of the imperfection of the geological record will rightly reject the whole theory. For he may ask where are the numberless transitional links which must have formerly connected the species? He may ask where are the remains of those infinitely numerous organisms which must have existed long before the Cambrian system was deposited? I can answer this last question only by supposing that, before the Cambrian epoch, the world presented a widely different aspect, and that the older continents exist now only as remnants in a metamorphosed condition, or lie still buried under the ocean.

Passing from these difficulties, the other leading facts of palaeontology agree admirably with the theory of descent with modification through variation and natural selection. We can thus understand how it is that new species come in slowly and successively, how species of different classes do not necessarily change together, yet in the long run all undergo modification to some extent. The extinction of old forms is the almost inevitable consequence of the production of new forms. We can understand why, when a species has once disappeared, it never reappears, for the link of generation has been broken.

We can understand how it is that the dominant forms which spread widely and yield the greatest number of varieties tend to people the world with allied but modified descendants, and how these will generally succeed in displacing the groups which are their inferiors in the struggle for existence. Hence after long periods of time the productions of the world appear to have changed simultaneously.

We can understand how it is that all the forms of life, ancient and recent, make together a few grand classes. We can understand, from the continued tendency to divergence of character, why the more ancient a form is the more it generally differs from those now living; why ancient and extinct forms often tend to fill up gaps between existing forms, for the more ancient a form is the more nearly it will be related to the common progenitor of groups since become widely divergent. We can see clearly why the organic remains of closely consecutive formations are closely allied.

The inhabitants of the world at each successive period in its history have beaten their predecessors in the race for life, and this may account for the belief held by many palaeontologists, that organisation on the whole has progressed. The succession of the same types of structure within the same area during the later geological periods ceases to be mysterious, and is intelligible on the principle of inheritance.

If then the geological record be as imperfect as many believe, the main objections to the theory of natural selection are greatly diminished or disappear. On the other hand, all the chief laws of palaeontology proclaim, as it seems to me, that species have been produced by ordinary generation: old forms having been supplanted by new and improved forms of life, the products of Variation and the Survival of the Fittest.

Present-day edentates: a hairy armadillo *(above)* **a two-toed sloth** *(below left)* **and a giant anteater** *(below right).*

Chapter 12
Geographical Distribution

In considering the distribution of organic beings over the face of the globe, the first great fact which strikes us is that neither the similarity nor the dissimilarity of the inhabitants of various regions can be wholly accounted for by climatal and other physical conditions. If we compare Australia, South Africa and western South America between latitudes 25° and 35°, we shall find parts extremely similar in all their conditions, yet it would not be possible to point out three faunas and floras more utterly dissimilar. Or again, we may compare the productions of South America south of latitude 35° with those north of 25°, which are exposed to considerably different conditions, yet they are incomparably more closely related to each other than they are to the productions of Australia or Africa under nearly the same climate.

A second great fact is that barriers to free migration are closely related to the differences between the productions of various regions. We see this in the great difference in nearly all the terrestrial productions of the New and Old Worlds, excepting in the northern parts, where the land almost joins and where, under a slightly different climate, there might have been free migration for the northern temperate forms. We see the same fact in the great difference between the inhabitants of Australia, Africa and South America under the same latitude; for these countries are almost as much isolated from each other as is possible. On each continent, also, we see the same fact; for on the opposite sides of lofty mountain-ranges, great deserts, and even large rivers, we find different productions; though as mountain-chains, etc.,

are not as impassable or likely to have endured so long as the oceans, the differences are very inferior to those characteristic of distinct continents.

Turning to the sea, we find the same law. The marine inhabitants of the eastern and western shores of South America are very distinct; but about thirty per cent of the fishes are the same on the opposite sides of the isthmus of Panama; and this fact has led naturalists to believe that the isthmus was formerly open. Westward of America, open ocean extends with not an island as a halting-place for emigrants; as soon as this barrier is passed we meet in the eastern islands of the Pacific with another and totally distinct fauna. So three marine faunas range far northward and southward in parallel lines not far from each other under corresponding climates; but from being separated by impassable barriers they are almost wholly distinct.

On the other hand, proceeding still farther westward from the eastern islands of the Pacific, we have innumerable islands as halting-places or continuous coasts until we come to the shores of Africa; and over this vast space we meet with no distinct marine faunas. Although so few marine animals are common to the three faunas of Eastern and Western America and the eastern Pacific islands, yet many shells are common to the eastern islands of the Pacific and the eastern shores of Africa on almost exactly opposite meridians of longitude.

A third great fact is the affinity of the productions of the same continent or sea. The naturalist, in travelling from north to south, never fails to be struck by the manner in which successive groups of beings specifically distinct, though nearly related, replace each other. He hears from closely allied yet distinct kinds of birds, notes nearly similar, and sees their nests similarly constructed, but not quite alike, with eggs coloured in nearly the same manner. The plains near the Straits of Magellan are inhabited by one species of *Rhea*, and northward the plains of La Plata by another species of the same genus, and not by an ostrich or emu like those inhabiting Africa and Australia under the same latitude. On these same plains we see the agouti and bizcacha, rodents having nearly the same habits as our hares and rabbits but displaying an American type of structure. We ascend the Cordillera and find an alpine species of bizcacha; we look to the waters and we do not find the beaver or musk-rat, but the coypu and capybara, rodents of the South American type. If we look to the American islands the inhabitants are essentially American, though they may be all peculiar species. We may look back to past ages, and we find American types then prevailing on the American continent and in the American seas. We see in these facts some deep organic bond independent of physical conditions. The naturalist must be dull who is not led to inquire what this bond is.

The bond is simply inheritance, which alone produces organisms like each other. The dissimilarity of the inhabitants of different regions may be attributed to modification through variation and natural selection. The degrees of dissimilarity will depend on the migration of the more dominant forms having been more or less effectually prevented at periods more or less remote, and especially on the relation of organism to organism in the struggle for life. Thus the high importance of barriers comes into play by checking migration, as does time for the slow process of modification through natural selection. Widely ranging species, numerous and already triumphant over many competitors, will have the best chance of seizing on new places when they spread into new countries. In their new homes they will be exposed to new conditions, and will frequently undergo further modification and produce groups of modified descendants. Thus we can understand how it is that genera are confined to the same areas, as is so commonly the case.

There is no evidence of the existence of any law of necessary development. As the variability of each species is an independent property, and will be taken advantage of only so far as it profits each individual, so the amount of modification in different species will be no uniform quantity. If a number of species, after having long competed with each other in their old home, were to migrate in a body into a new and afterwards isolated country, they would be little liable to modification; for neither migration nor isolation in themselves effect anything. These principles come into play only by bringing organisms into new relations with each other, and in a lesser degree with the surrounding physical conditions. As we have seen in the last chapter that some forms have retained nearly the same character from an enormously remote geological period, so certain species have migrated over vast spaces and have not become greatly or at all modified.

Single Centres of supposed Creation

We are thus brought to the question which has been largely discussed by naturalists, namely, whether species have been created at one or more points of the earth's surface. Undoubtedly there are many cases of extreme difficulty in understanding how the same species could possibly have migrated from some one point to the several distant and isolated points where now found. Nevertheless, the simplicity of the view that each species was first produced within a single region captivates the mind. He who rejects it rejects the *vera causa* of ordinary generation with subsequent migration, and calls in the agency of a miracle. It is universally admitted that in most cases the area inhabited by a species is continuous, so that when a plant or animal inhabits two points so distant that the space could not have been easily passed over by migration, the fact is given as something exceptional.

The incapacity of migrating across a wide sea is more clear with terrestrial mammals than with any other beings; and we find no inexplicable instances of the same mammals inhabiting distant points. No geologist feels any difficulty in Britain possessing the same quadrupeds with the rest of Europe, for they were once united. But if the same species can be produced at two separate points, why do we not find a single mammal common to Europe and Australia or South America? The conditions of life are nearly the same, so that European animals and plants have become naturalised in America and Australia; and some of the aboriginal plants are identical at these distant points.

The answer is that mammals have not been able to migrate, whereas some plants, from their varied means of dispersal, have migrated across the wide interspaces. The great influence of barriers is intelligible only on the view that the great majority of species have been produced on one side, and have not been able to migrate to the opposite side. Some few families, very many genera, and a still greater number of sections of genera are confined to a single region.

Hence it seems to me that the view of each species having been produced in one area alone, and having subsequently migrated, is the most probable. Undoubtedly many cases occur in which we cannot explain how the same species could have passed from one point to the other. But geographical and climatal changes which have occurred within recent geological times have rendered discontinuous the formerly continuous range of many species. So we are reduced to consider whether the exceptions are so numerous and of so grave a nature that we ought to give up this belief. Accordingly, after some preliminary remarks, I will discuss a few of the most striking exceptions; namely, the existence of the same species on the summits of distant mountain-ranges, and at distant points in the arctic and antarctic regions; and (in the following chapter), the wide distribution of fresh-water productions, and the occurrence of the same terrestrial species on islands and on the nearest mainland, though separated by hundreds of miles of open sea. If the existence of the same species at distant points of the earth's surface can be explained on the view of each species having migrated from a single birthplace, then the belief that a single birthplace is the law seems to me incomparably the safest.

Before discussing the three classes of facts which I have selected as presenting the greatest amount of difficulty on the theory of 'single centres of creation', I must say a few words on the means of dispersal.

Means of Dispersal

I can give here only the more important facts. Change of climate must have had a powerful influence on migration. A region now impassable to certain organisms, from the nature of its climate, might have been a high road for migration when the climate was different. Changes of level in the land must also have been highly influential: a narrow isthmus now separates two marine faunas; submerge it, and the two faunas will blend together. Where the sea now extends, land may at a former period have connected islands or possibly even continents together, and thus have allowed terrestrial productions to pass from one to the other.

[*The continents were all once linked: see the maps of continental drift on pp. 188–9. Although this was before what Darwin calls the 'Recent period' (see glossary for this meaning of Recent) their former linkage has influenced the distribution of existing plants and animals. Darwin is right, however, in observing that most oceanic islands are the results of volcanic activity and have never been linked to a continent.*]

No geologist disputes that great mutations of level have occurred within the period of existing

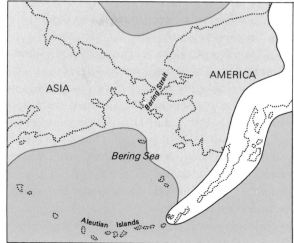

'Where the sea now extends, land may at a former period have connected islands or possibly even continents together, and thus have allowed terrestrial productions to pass from one to the other.' During the ice ages so much water was locked in the glaciers that the sea level fell by as much as 90 metres (290 feet) thus draining the Bering Strait of water and establishing a link between the Asian and American continents. A large number of mammals travelled across the Bering land bridge: horses, and the ancestors of the camel were among those that moved from America into Asia; mammoths and bison, together with many others, invaded America. Asiatic peoples also crossed into America and from them the Indians of both North and South America are descended.

organisms. Edward Forbes insisted that all the islands in the Atlantic must have been recently connected with Europe or Africa, and Europe likewise with America. This view removes many a difficulty; but to the best of my judgment we are not authorised in admitting such enormous geographical changes within the period of existing species. I freely admit the former existence of many islands, now buried beneath the sea, which may have served as halting-places for plants and animals during their migration. In the coral-producing oceans such sunken islands are now marked by rings of coral or atolls standing over them. But I do not believe that within the Recent period most of our continents have been united with each other and the existing oceanic islands.

The almost universally volcanic composition of such islands does not favour the admission that they are the wrecks of sunken continents.

I must now say a few words on occasional means of distribution. I shall here confine myself to plants. In botanical works, this or that plant is often stated to be ill-adapted for wide dissemination; but the greater or less facilities for transport across the sea may be said to be almost wholly unknown. Until I tried a few experiments, it was not even known how far seeds could resist the injurious action of sea-water. To my surprise I found that out of 87 kinds, 64 germinated after an immersion of 28 days, and a few survived 137 days. For convenience' sake I chiefly tried small seeds, and all of these sank in a few days. Afterwards I tried some larger fruits, capsules, etc., and some of these floated for a long time. It occurred to me that floods would often wash into the sea dried plants with seed-capsules or fruit attached to them. Hence I dried the stems and branches of 94 plants with ripe fruit and placed them on sea-water. The majority sank quickly, but 18 floated for above 28 days; and some of the 18 floated for a very much longer period. We may infer from these scanty facts that the seeds of $\frac{14}{100}$ kinds of plants of any country might float during 28 days and retain their power of germination. In Johnston's Physical Atlas the average rate of the Atlantic currents is 33 miles per diem; on this average, the seeds of $\frac{14}{100}$ plants belonging to one country might be floated across 924 miles of sea to another country, and if stranded would germinate.

Seeds may be occasionally transported in another manner. Drift-timber is thrown up on most islands, even in the midst of the widest oceans. When stones are embedded in the roots of trees, small parcels of earth are frequently enclosed in their interstices so perfectly that not a particle could be washed away during the longest transport: out of one small portion of earth thus *completely* enclosed by the roots of an oak, three dicotyledonous plants germinated. Again the carcasses of birds, when floating on the sea, sometimes escape being immediately devoured: and many kinds of seeds in the crops of floating birds long retain their vitality.

Living birds are highly effective agents in the transportation of seeds. Frequently birds are blown by gales vast distances across the ocean. Nutritious seeds do not pass through their intestines, but hard seeds pass uninjured through even the digestive organs of a turkey. The following fact is even more important: the crops of birds do not secrete gastric juice, and do not injure the germination of seeds; now, after a bird has devoured its food, all the grains do not pass into the gizzard for twelve or even eighteen hours. A bird in this interval might easily be blown 500 miles, and hawks are known to look out for tired birds, and the contents of their torn crops might thus readily get scattered. Some hawks bolt their prey whole, and after an interval disgorge pellets, which include seeds capable of germination. Some seeds of the oat, wheat, millet, hemp, and clover germinated after having been from twelve to twenty-one hours in the stomachs of different birds of prey; and two seeds of beet grew after having been thus retained for two days and fourteen hours.

Locusts are sometimes blown to great distances from the land; I myself caught one 370 miles from the coast of Africa and in November 1844 swarms of locusts visited the island of Madeira. Now in parts of Natal it is believed that injurious seeds are introduced in the dung left by great flights of locusts: out of such dried pellets I extracted several seeds, and raised from them seven grass plants belonging to two species of two genera. Hence a swarm of locusts might be the means of introducing several kinds of plants into an island lying far from the mainland.

Although the beaks and feet of birds are generally clean, earth sometimes adheres to them: the leg of a woodcock was sent me with a little cake of dry earth attached to the shank, weighing only nine grains; and this contained a seed of the toad-rush which germinated and flowered. Mr. Swaysland, who has paid close attention to our migratory birds, has often shot wagtails, wheatears and whinchats on their first arrival before they had alighted; he has several times noticed little cakes of earth attached to their feet, and generally soil is charged with seeds. Professor Newton sent me the leg of a red-legged partridge with a ball of hard earth adhering to it weighing six and half ounces. The earth had been kept for three years, but when broken, watered and placed under a bell glass, no less than 82 plants sprung from it. Can we doubt that the many birds which are annually blown across great spaces of ocean and which annually migrate must occasionally transport a few seeds embedded in dirt adhering to their feet or beaks?

As icebergs are sometimes loaded with earth and have even carried brushwood and the nest of a land-bird, they must occasionally have transported seeds in the arctic and antarctic regions and, during the Glacial period, within the now temperate regions. In the Azores, from the large number of plants common to Europe in comparison with the other islands of the Atlantic, and from their somewhat northern character in comparison with the latitude, I suspect that these islands have been partly stocked by ice-borne seeds during the Glacial epoch. As M. Hartung has observed on these islands erratic boulders of granite and other rocks which do not occur in the archipelago, we may infer that icebergs formerly landed their burdens on the shores of these mid-ocean islands, and it is at least possible that they may have brought thither some few seeds of northern plants.

Considering that these and other means of transport have been in action year after year for tens of thousands of years, it would be a marvellous fact if many plants had not thus become widely transported. It should be observed that scarcely any means would carry seeds for very great distances: for seeds do not retain their vitality when exposed for a great length of time to sea-water; nor could they be long carried in the crops or intestines of birds. These means, however, would suffice for occasional transport across some hundred miles or from island to island, or from a continent to a neighbouring island, but not from one distant continent to another. The floras of distant continents would not by such means become mingled; but would remain as distinct as they now are. The currents, from their course, would never bring seeds from North America to Britain. Almost every year one or two land-birds are blown from North America to Ireland and England; but seeds could be transported by these rare wanderers only by dirt adhering to their feet or beaks. Even in this case, how small would be the chance of a seed falling on favourable soil and coming to maturity on these well-stocked islands! But a poorly stocked island, remote from the mainland, might receive colonists by such means, and before the island had become fully stocked with inhabitants, with few destructive insects or birds living there, nearly every seed which chanced to arrive, if fitted for the climate, would germinate and survive.

Dispersal during the Glacial Period

The identity of many plants and animals on mountain-summits separated by hundreds of miles of lowlands where Alpine species could not possibly exist, is one of the most striking cases of the same species living at distant points without the apparent possibility of their having migrated. It is indeed remarkable to see so many plants of the same species living on the snowy regions of the Alps or Pyrenees, and in the extreme northern parts of Europe; but it is far more remarkable that the plants on the White Mountains in the United States are the same with those of Labrador, and nearly all the same with those on the loftiest mountains of Europe.

But the Glacial period affords a simple explanation of these facts. We have evidence of almost every conceivable kind that, within a very recent geological period, Central Europe and North America suffered under an arctic climate. The ruins of a house burnt by fire do not tell their tale more plainly than do the mountains of Scotland and Wales, with their scored flanks, polished surfaces, and perched boulders, of the icy streams with which their valleys were lately filled.

We shall follow the influence of glaciation by supposing a new glacial period slowly to come on, and then pass away. As the cold came on, the inhabitants of the north would take the places of the former inhabitants of the temperate regions, and the latter would travel farther southward or perish. The mountains would become covered with snow and ice, and their former Alpine in-

habitants would descend to the plains. By the time the cold had reached its maximum, arctic fauna and flora would cover Europe as far as the Alps and Pyrenees, and the now temperate regions of the United States, whose plants and animals would be nearly the same with those of Europe; for the present circumpolar inhabitants, which we suppose to have travelled southward, are remarkably uniform round the world.

As the warmth returned, the arctic forms would retreat northward, or ascend the mountains as the snow disappeared higher and higher. Hence, when the warmth had fully returned, the same species would be found in the arctic regions of the Old and New Worlds and on many isolated mountain-summits.

Thus we can understand the identity of many plants at points so immensely remote as the mountains of the United States and those of Europe. We can thus also understand the fact that the Alpine plants of each mountain-range are more especially related to the arctic forms living due north or nearly due north of them: for the first migration when the cold came on, and the remigration on the returning warmth, would generally have been due south and north.

As the arctic forms migrated in a body together, their mutual relations will not have been much disturbed. Hence these forms will not have been liable to much modification. But with the productions isolated on the mountains by the returning warmth, the case will have been different, for it

(above left) **Ferns colonizing volcanic rock on Hawaii. Although the chances of a seed being transported across hundreds or thousands of miles of ocean to a volcanic island are relatively small, those that do get there stand a good chance of establishing themselves since there is little competition from other plants for growing space and light and few or no destructive animals.**

(opposite) **Mountain avens, a plant adapted to cold climates, growing in the European Alps. This plant is also found in the extreme north of Europe, but not in the intervening regions. It probably originated in Northern Europe, moved southwards during the ice ages, and then, when the glaciers retreated, survived only in the mountains.**

is not likely that all the same arctic species will have been left on mountain-ranges far distant from each other, and they will probably have mingled with ancient Alpine species, which must have existed before the Glacial epoch; they will, also, have been subsequently exposed to different climatal influences. Their mutual relations will thus have been disturbed; consequently they will have been liable to modification; and if we compare the present Alpine plants and animals of the great European mountain-ranges, though many of the species remain the same, some exist as varieties, and some as closely allied species representing each other on the several ranges.

Believing that our continents have long remained in nearly the same relative position, though subjected to great oscillations of level, I am strongly inclined to infer that during some earlier and warmer period, such as the Pliocene, a large number of the same plants and animals inhabited the almost continuous circumpolar land; these plants and animals, both in the Old and New Worlds, migrated southwards as the climate became less warm, long before the Glacial period. We now see their descendants, mostly in a modified condition, in Europe and the United States. On this view we can understand the relationship between the productions of North America and Europe, and the fact that during the later Tertiary stages they were more closely related to each other than at present; for during these warmer periods the northern parts of the Old and New Worlds will have been almost continuously united by land, serving as a bridge, since rendered impassable by cold, for the intermigration of their inhabitants.

During the slowly decreasing warmth of the Pliocene period, as soon as species migrated south of the Polar Circle they will have been completely cut off from each other. This separation, as far as the more temperate productions are concerned, must have taken place long ages ago. As they migrated southward, they will have mingled in

the one region with the native American productions, and in the other with those of the Old World. Consequently everything favours far more modification than with the Alpine productions, isolated on mountain-ranges. We find very few identical species among the temperate productions of the New and Old Worlds, but we find in every great class many closely allied or representative forms.

Alternate Glacial Periods in the North and South

[*These facts about plant distribution can now be explained on the basis of the former unity of the continents in one land mass (see maps on pp. 188–9). Slight variations in the inclination and orientation of the earth's rotational axis are now thought to be one of the factors which contribute to the intiation of Ice Ages. However, these variations are not large enough to have caused the glacial periods to occur alternately in the Northern and Southern Hemispheres.*]

According to Mr. Croll, whenever the northern hemisphere passes through a cold period, the temperature of the southern hemisphere is actually raised. So conversely it will be with the northern hemisphere whilst the southern passes through a glacial period. This conclusion throws much

(left) '*The ruins of a house burnt by fire do not tell their tale more plainly than do the mountains of Scotland and Wales of the icy streams with which their valleys were lately filled.*'

(overleaf) **Only very recently has the theory of continental drift become established. Although the close fit between the continents, particularly the east coast of South America and the west coast of Africa, had been noted since the seventeenth century, it seemed impossible for continents to move. Discovery of the spreading of the sea floor at the mid-oceanic ridges led to the formulation of the concept of the surface of the earth being made up of moving plates of oceanic crust on which the continents are carried. This is the theory of plate tectonics, which has led to the acceptance of continental drift. Several of Darwin's observations can be explained on the basis of continental drift, for example the similarity of plants and lower organisms from different continents, and the dissimilarity of the mammals which largely evolved after the continents had split up. The maps show the areas of land (brown), continental shelf (blue-green) and sea (blue) at six points in the history of the earth. On each map the positions of the present-day land masses are depicted by a black line.**

1. The Cambrian; 510 million years ago. At this time there were four large continents, North America, Europe, Gondwanaland (South America, Antarctica, Africa and Australia) and Angara (eastern Asia). Europe and North America were moving towards each other.

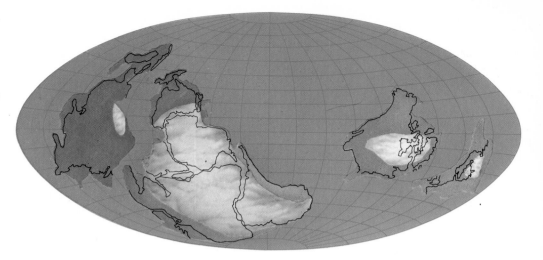

2. The Devonian; 380 million years ago. Europe and North America had collided so that there were only three continents. Sea levels were low and there were recurrent periods of drought in some regions.

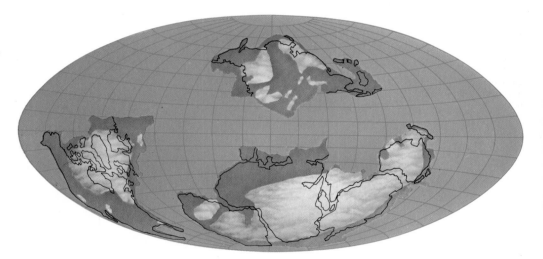

3. The Permian; 250 million years ago. All three continents had collided to form one supercontinent, known as Pangaea. Again sea levels were low and desert conditions widespread, but in between the Devonian and the Permian there was a period of great humidity, the Carboniferous, when swamps and forests dominated the earth.

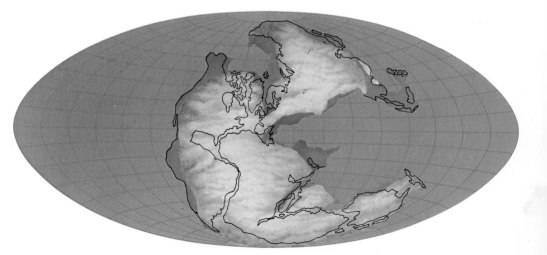

4. The Cretaceous; 100 million years ago. North America and Europe had begun to split apart, and the rift between South America and Africa followed shortly afterwards. The sea level was very high.

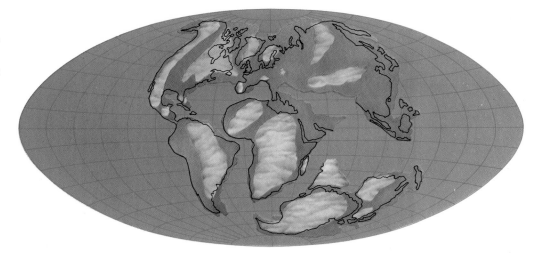

5. The Eocene; 50 million years ago. The Atlantic Ocean had opened up. South America was an island, and so was India which had broken away from Africa and was moving northwards. Australia was still locked on to Antarctica.

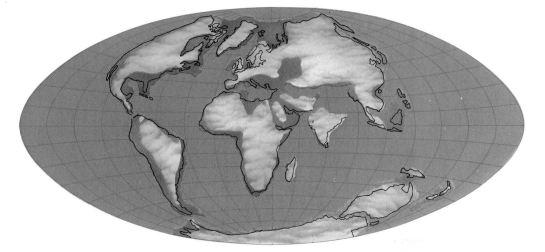

6. The Pleistocene; 40,000 years ago. The continents were almost in their present-day positions. During the Pleistocene there were four major glaciations. During these the sea level was very low, so that the coastline was near the edge of the continental shelf. In between the glaciations there were very warm periods.

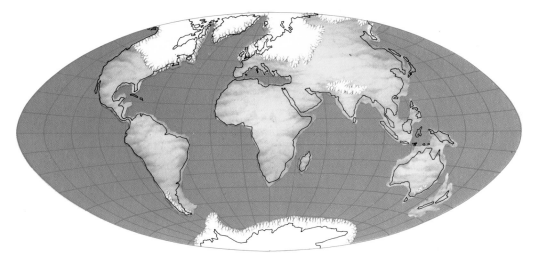

The ruins at Concepción after the earthquake of 1835, sketched by J.C. Wickham, First Lieutenant of the *Beagle*. Darwin was ashore collecting fossils when the earthquake struck, and was shocked by the force of the tremors. How surprised he would have been to know that the earthquake was an outcome of the same process that caused the strange similarities Joseph Hooker reported between the plants of South America, South Africa and Australia: continental drift. These similarities in flora stem from the time when the southern continents were united as Gondwanaland (see maps on pp.188-9). The north-south split that opened up into the Atlantic Ocean is still widening by a few centimetres every year. As it does so the plates underlying the Pacific are forced against the plates on which the Americas are carried. The Pacific plate slides underneath and is absorbed back into the earth, but this movement causes the frequent earthquakes of the west coast of the Americas, as well as pushing up mountain chains such as the Rockies and the Andes.

light on geographical distribution; but I will first give the facts which demand an explanation.

In South America, between forty and fifty of the flowering plants of Tierra del Fuego, forming no inconsiderable part of its scanty flora, are common to North America and Europe. On the Organ Mountains of Brazil, some few temperate European, some Antarctic, and some Andean genera were found by Gardner which do not exist in the low intervening hot countries. In Africa, several forms characteristic of Europe and the Cape of Good Hope occur on the mountains of Abyssinia. Several of the plants living on the upper parts of the lofty island of Fernando Po and on the neighbouring Cameroon Mountains are closely related to those on the mountains of Abyssinia and those of temperate Europe. It now also appears that some of these same temperate plants have been discovered on the mountains of the Cape Verde Islands. This extension of the same temperate forms, almost under the equator, across the whole continent of Africa and to the mountains of the Cape Verde Archipelago is one of the most astonishing facts ever recorded in the distribution of plants.

On the Himalaya and on the isolated mountain-ranges of the peninsula of India, on the heights of Ceylon and on the volcanic cones of Java, many plants occur either identically the same or representing each other and at the same time representing plants of Europe not found in the intervening hot lowlands. Still more striking is the fact that peculiar Australian forms are represented on the summits of Borneo. On the southern mountains of Australia Dr. F. Müller has discovered several European species; other species, not introduced by man, occur on the lowlands; and a long list can be given, as I am informed by Dr. Hooker, of European genera found in Australia but not in the intermediate torrid regions. Hence we see that certain plants growing on the more lofty mountains of the tropics in all parts of the world and on the temperate plains of the north and south are either the same species or varieties of the same species. Besides these forms, many species inhabiting the same widely sundered areas belong to genera not now found in the intermediate tropical lowlands.

Now let us see whether Mr. Croll's conclusion that when the northern hemisphere suffered from the great Glacial period, the southern hemisphere was actually warmer, throws any light on this present apparently inexplicable distribution of organisms. We know that during the Glacial period Arctic forms invaded the temperate regions; and some of the temperate forms invaded the equatorial lowlands. The inhabitants of these hot lowlands would at the same time have migrated to the south, for the southern hemisphere was at this period warmer. On the decline of the Glacial period, as both hemispheres recovered their former temperatures, the northern temperate forms living on the lowlands under the equator would have been driven to their former homes, or would have ascended adjoining high land where they would have long survived like the Arctic forms on the mountains of Europe.

The southern hemisphere would in its turn be subjected to a severe Glacial period, with the northern hemisphere rendered warmer, and then the southern temperate forms would invade the equatorial lowlands. The northern forms which had been left on the mountains would now descend and mingle with the southern forms. These latter, when the warmth returned, would return to their former homes, leaving some few species on the mountains and carrying southward some of the northern temperate forms which had descended from the mountains.

I am far from supposing that all the difficulties in regard to the distribution of the identical and allied species which now live so widely separated in the north and south and on the intermediate mountain-ranges, are removed on the views above given. The exact lines of migration cannot be indicated. We cannot say why certain species and not others have migrated; why certain species have given rise to new forms, whilst others have remained unaltered. We cannot hope to explain such facts until we can say why one species and not another becomes naturalised by man in a foreign land; why one species ranges twice as far and is twice as common as another species within their own homes.

Various special difficulties also remain to be solved; for instance the occurrence, as shown by Dr. Hooker, of the same plants at points so enormously remote as Kerguelen Land, New Zealand, and Fuegia; but icebergs, as suggested by Lyell, may have been concerned in their dispersal. The existence at these and other distant points of the southern hemisphere of species which, though distinct, belong to genera exclusively confined to the south, is a more remarkable case. Some of these species are so distinct that we cannot suppose that there has been time since the commencement of the last Glacial period for their migration and subsequent modification to the necessary degree.

The facts seem to indicate that distinct species belonging to the same genera have migrated in radiating lines from a common centre; and I am inclined to look in the southern, as in the northern hemisphere, to a former and warmer period before the last Glacial period, when the Antarctic lands, now covered with ice, supported a highly peculiar and isolated flora. It may be suspected that before this flora was exterminated a few forms had been already widely dispersed to various

The plants of the family Proteaceae are found only in
Australia, southwest Asia, southern Africa, south and central
America. Within this family, which is not closely related to any
other plant family, there are several examples of the links
between the floras of Australia, South Africa and South
America which Joseph Hooker observed and which Darwin
correctly predicted would *'some day be explained'*. One genus,
for example, has three species of which one is native to
Australia, one to New Guinea and one to Chile. Shown here
are plants of the Proteaceae family from Tierra del Fuego *(left)*,
South Africa *(above)* and Australia *(right)*.

points of the southern hemisphere by occasional
means of transport and by the aid, as halting-
places, of now sunken islands. Thus the southern
shores of America, Australia and New Zealand
may have become slightly tinted by the same
peculiar forms of life. The affinity which, though
feeble, I am assured by Dr. Hooker is real, between
the flora of the south-western corner of Australia
and of the Cape of Good Hope, is a far more remarkable case but will, no doubt, some day be
explained.

Chapter 13

Geographical Distribution
(continued)

Fresh-water Productions

As lakes and river-systems are separated by barriers of land and sea, it might have been thought that fresh-water productions would not have ranged widely. But the case is exactly the reverse. Not only have many fresh-water species an enormous range, but allied species prevail throughout the world. When first collecting in the fresh waters of Brazil I well remember feeling much surprise at the similarity of the fresh-water insects, shells, etc., and at the dissimilarity of the surrounding terrestrial beings, compared with those of Britain.

But the wide ranging power of fresh-water productions can in most cases be explained by their having become fitted for short and frequent migrations from pond to pond within their own countries. We can here consider only a few cases. It was formerly believed that the same fresh-water species of fish never existed on two continents distant from each other. But Dr. Günther has lately shown that the *Galaxias attenuatus* inhabits Tasmania, New Zealand, the Falkland Islands, and South America. This probably indicates dispersal from an Antarctic centre during a former warm period. This case, however, is rendered less surprising by the species of this genus having the power of crossing considerable spaces of open ocean: thus there is one species common to New Zealand and to the Auckland Islands, separated by 230 miles. On the same continent fresh-water fish often range widely, and as if capriciously; for in two adjoining river-systems some of the species may be the same and some wholly different. Their dispersal may be mainly attributed to changes in the level of the land within the recent period, causing rivers to flow into each other. Instances, also, could be given of this having occurred during floods, without any change of level.

The wide difference of the fish on the opposite sides of continuous mountain-ranges, which from an early period have prevented the inosculation of the river-systems on the two sides, leads to the same conclusion. Some fresh-water fish belong to very ancient forms, and in such cases there will have been ample time for great geographical changes and much migration. Salt-water fish can be slowly accustomed to live in fresh water; and there is hardly a single group of which all the members are confined to fresh-water, so that a marine species belonging to a fresh-water group might travel far along the shores of the sea, and could become adapted to the fresh waters of a distant land.

Some species of fresh-water shells have very wide ranges, and allied species prevail throughout the world. Their distribution at first perplexed me much, as their ova and adults are immediately killed by sea-water. I could not even understand how species have spread rapidly throughout the same country. But two facts throw some light on this subject. When ducks suddenly emerge from a pond covered with duck-weed, I have twice seen these little plants adhering to their backs; and it has happened to me, in removing a little duck-weed from one aquarium to another, that I have unintentionally stocked the one with shells from the other. But another agency is perhaps more effectual: I suspended the feet of a duck in an aquarium, where many ova of fresh-water shells were hatching; and I found that numbers of the extremely minute and just-hatched shells crawled on the feet and clung to them so firmly that they could not be jarred off, though at a somewhat more advanced age they would voluntarily drop off. These molluscs, though aquatic in their nature, survived on the duck's feet, in damp air, from twelve to twenty hours; and in this time a duck might fly at least six hundred miles, and if blown across the sea to any distant point would be sure to alight on a pool or rivulet. Sir Charles Lyell informs me that a *Dytiscus* [diving-beetle] has been caught with an *Ancylus* (a fresh-water shell like a limpet) firmly adhering to it; and a water-beetle of the same family, a *Colymbetes*, once flew on board the *Beagle* when forty-five miles from land: how much farther it might have been blown by a favouring gale no one can tell.

With respect to plants, many fresh-water species have enormous range both over continents and to the most remote oceanic islands. I think favourable means of dispersal explain this fact. I have before mentioned that earth occasionally adheres to the feet and beaks of birds. Wading-birds wander more than others, and they are occasionally found on the most remote islands of the ocean; they would not be likely to alight on the sea, so that, when gaining land, they would be sure to fly to their natural fresh-water haunts. I do not believe that botanists are aware how charged the mud of ponds is with seeds: I took in February three tablespoonfuls of mud from a little pond: this mud when dried weighed only $6\frac{3}{4}$ ounces; I kept it covered up for six months, pulling up and counting each plant as it grew; the plants were of many kinds, and were altogether 537 in number; and yet the viscid mud was all contained in a breakfast cup! Considering these facts, I think it would be an inexplicable circumstance if water-birds did not transport the seeds of fresh-water plants to ponds at very distant points. The same agency may have come into play with the eggs of smaller fresh-water animals.

Other and unknown agencies have also played a part. Fresh-water fish eat some kinds of seeds; as of the yellow water-lily. Herons and other birds devour fish; they then take flight and go to other waters; and seeds retain their power of germination when rejected many hours afterwards in pellets or in the excrement. Audubon states that he found the seeds of the great southern water-lily (probably, according to Dr. Hooker, the *Nelumbium luteum*) in a heron's stomach. Now this bird must often have flown with its stomach thus well stocked to distant ponds, and then getting a hearty meal of fish, analogy makes me believe that it would have rejected the seeds in a pellet in a fit state for germination.

In considering these several means of distribution, it should be remembered that when a pond or stream is first formed, for instance on a rising islet, it will be unoccupied; and a single seed or egg will have a good chance of succeeding. We should also not forget the probability of many fresh-water forms having formerly ranged continuously over immense areas, and then having become extinct at intermediate points. But the wide distribution of fresh-water plants and the lower animals apparently depends in main part on the wide dispersal of their seeds and eggs by animals, especially birds which naturally travel from one piece of water to another.

On the Inhabitants of Oceanic Islands

I have already given my reasons for disbelieving in continental extensions within the period of existing species, on so enormous a scale that all the islands of the oceans were thus stocked with their present terrestrial inhabitants. This view removes many difficulties, but does not accord with all the facts in regard to the productions of islands.

The species which inhabit oceanic islands are few compared with those on equal continental areas: New Zealand, for instance, together with the outlying islands of Auckland, Campbell and Chatham, contains only 960 kinds of flowering plants; if we compare this moderate number with the species which swarm over equal areas in South-western Australia or the Cape of Good Hope, we must admit that some cause, independently of different physical conditions, has given rise to so great a difference in number. Even the uniform county of Cambridge has 847 plants, but a few ferns and a few introduced plants are included in this number, and the comparison in some other respects is not quite fair. We have evidence that the barren island of Ascension aboriginally possessed less than half a dozen flowering plants; yet many species have now become naturalised on it, as they have in New Zealand and every other oceanic island. He who admits the doctrine of the creation of each separate species will have to admit that a sufficient number of the best adapted plants and animals were not created for oceanic islands; for man has unintentionally stocked them far more fully and perfectly than did nature.

Although in oceanic islands the species are few in number, the proportion of endemic kinds (i.e. those found nowhere else) is often extremely large. If we compare, for instance, the number of endemic land-shells in Madeira, or of endemic birds in the Galapagos Archipelago, with the number on any continent, and then compare the area of the island with that of the continent, we shall see that this is true.

This might have been expected, for species arriving in an isolated district and having to compete with new associates would be eminently liable to modification, and would often produce modified descendants. But it by no means follows that the species of another class or another section of the same class would be peculiar. In the Gala-pagos Islands, of the 26 land-birds 21 are peculiar, whereas of the 11 marine birds only 2 are peculiar; it is obvious that marine birds could arrive at these islands much more easily and frequently than land-birds, so that the insular forms would inter-cross with unmodified immigrants from the mother-country. Bermuda, on the other hand,

which lies at the same distance from America as the Galapagos Islands, does not possess a single endemic land-bird; and we know that very many North American birds frequently visit this island. So it has been stocked from the neighbouring continent with birds, which for long ages have struggled together and become mutually co-adapted. Hence when settled in their new homes each kind will have been kept by the others to its proper place, and will consequently have been

The islands discussed by Darwin, and the major ocean currents. Darwin noted that, with only a few exceptions, the inhabitants of islands were related to those of the nearest mainland. One exception is Kerguelen whose plants Joseph Hooker observed to be closely related to those of South America. This can be understood on the basis that this island has been stocked with seeds carried by the prevailing currents.

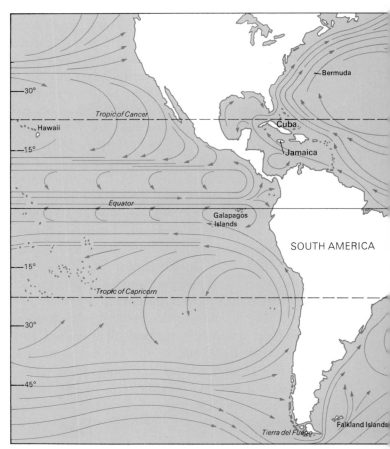

but little liable to modification. Any tendency to modification will also have been checked by intercrossing with unmodified immigrants, often arriving from the mother-country. Madeira is inhabited by a wonderful number of peculiar land-shells, whereas not one species of sea-shell is peculiar to its shores: now we can see that the eggs or larvae of sea-shells, perhaps attached to seaweed or floating timber, might be transported across three or four hundred miles of sea far more easily than land-shells.

Oceanic islands are sometimes deficient in animals of certain whole classes, and their places are occupied by other classes: thus in the Galapagos Islands reptiles, and in New Zealand gigantic wingless birds, take or recently took the place of mammals. Turning to plants, in the Galapagos Islands the proportional numbers of the different orders are very different from what they are elsewhere. All such differences are generally accounted for by supposed differences in the physical conditions, but this explanation is not a little doubtful. Facility of immigration seems to have been fully as important as the nature of the conditions.

In certain islands not tenanted by a single mammal, some of the endemic plants have beautifully hooked seeds; yet hooks undoubtedly serve for the transportal of seeds in the fur of quadrupeds. But a hooked seed might be carried to an island by other means; and the plant then becoming modified would form an endemic species, still retaining

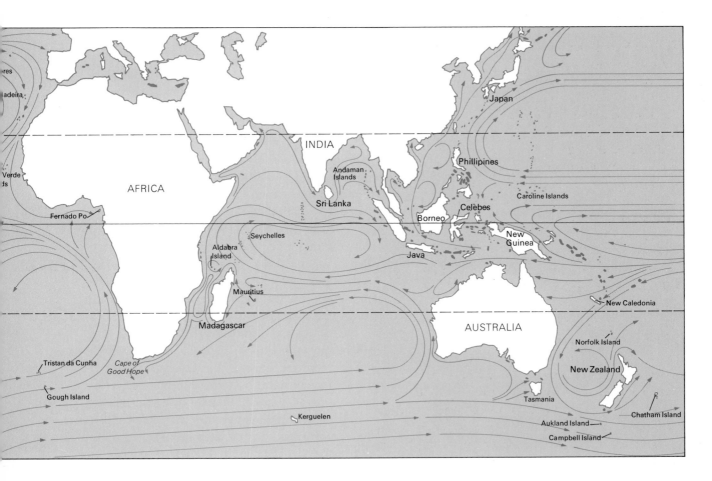

its hooks, which would form a useless appendage like the shrivelled wings of many insular beetles.

Absence of Batrachians [Amphibians] and Terrestrial Mammals on Oceanic Islands

With respect to the absence of whole orders, Batrachians (frogs, toads, newts) are never found on any of the islands with which the great oceans are studded, with the exception of New Zealand, New Caledonia, the Andaman Islands, and perhaps the Salomon Islands and the Seychelles. But it is doubtful whether New Zealand and New Caledonia ought to be classed as oceanic islands; and this is still more doubtful with respect to the Andaman and Salomon groups and the Seychelles. This general absence of frogs, toads and newts on true oceanic islands cannot be accounted for by their physical conditions: for frogs have been introduced into Madeira, the Azores, and Mauritius, and have multiplied so as to become a nuisance. But as these animals and their spawn are immediately killed by sea-water, there would be great difficulty in their transportal across the sea.

Mammals offer a similar case. I have not found a single instance free from doubt of a terrestrial mammal (excluding domesticated animals) inhabiting an island situated above 300 miles from a continent or great continental island; and many islands at a much less distance are equally barren. The Falkland Islands, which are inhabited by a wolf-like fox, come nearest to an exception; but this group lies on a bank in connection with the mainland 280 miles distant; moreover, icebergs may have formerly transported foxes, as now frequently happens in the arctic.

It cannot be said that there has not been time for the creation of mammals; many volcanic islands are sufficiently ancient; there has also been time for the production of endemic species belonging to other classes; and on continents new species of mammals appear and disappear at a quicker rate than other animals. Although terrestrial mammals do not occur on oceanic islands, aerial mammals occur on almost every island: New Zealand possesses two bats found nowhere else in the world: Norfolk Island, the Viti Archipelago, the Bonin Islands, the Caroline and Marianne Archipelagos and Mauritius all possess their peculiar bats. Why has the supposed creative force produced bats and no other mammals on remote islands? On my view this question can easily be answered; for no terrestrial mammal can be transported across a wide sea, but bats can fly across. Bats have been seen wandering by day far over the Atlantic Ocean; many species have enormous ranges and are found on continents and far distant islands.

Another interesting relation exists, namely between the depth of the sea separating islands from each other or from the nearest continent, and the degree of affinity of their mammalian inhabitants. Mr. Windsor Earl has made some striking observations on this head, since greatly extended by Mr. Wallace's admirable researches, in regard to the great Malay Archipelago, which is traversed near Celebes by deep ocean separating two widely distinct mammalian faunas. On either side the islands stand on a moderately shallow bank and are inhabited by the same or closely allied quadrupeds. As the amount of modification which animals undergo partly depends on time, and as the islands which are separated by shallow channels are more likely to have been united within a recent period than islands separated by deeper channels, we can understand how it is that a relation exists between the depth of sea separating two mammalian faunas and the degree of their affinity.

The foregoing statements in regard to the inhabitants of oceanic islands accord better with the belief in the efficiency of occasional means of transport than with the belief in the former connection of all oceanic islands with the nearest continent; for on this latter view it is probable that the various classes would have immigrated more uniformly, that their mutual relations would not have been much disturbed, and that consequently they would not have been much modified.

On the Relations of the Inhabitants of Islands to those of the nearest Mainland

The most striking fact is the affinity of species inhabiting islands to those of the nearest mainland, without being actually the same. The Galapagos Archipelago lies between 500 and 600 miles

Oceanic islands have a unique natural history largely because certain groups of animals tend to colonize them, while others are unable to do so. Flying animals, that is insects, birds and bats, can reach the island. Seabirds fly there and leave again with relative ease: they may become adapted to life on land, as the flightless cormorant (p.96) has done, or, like the frigate bird (p.80), they may treat the island only as a stopping place. Smaller birds, insects and bats are generally blown to the island by accident, and must adapt to life there or perish. Reptiles probably reach islands by floating on driftwood; they can survive the long periods without food and water necessary for this journey. Mammals cannot do this because they have a higher metabolism and non-flying land mammals do not colonize islands unless these are very close to the mainland. Amphibians cannot live away from freshwater, so they too are excluded. Sea mammals, such as seals and sea lions, have little difficulty in reaching an island, but return to the sea after the breeding season and do not colonize the interior. Crabs, on the other hand, often evolve novel forms on islands; they may stay near the shoreline, as with the Sally-lightfoot crabs of the Galapagos, or adapt to an inland niche.

from South America. Here almost every product of the land and water bears the unmistakable stamp of the American continent. Why should this be so? Why should the species which are supposed to have been created in the Galapagos Archipelago and nowhere else bear so plainly the stamp of affinity to those created in America? There is nothing in the conditions of life, in the geological nature of the islands or their climate, which closely resembles the conditions of the South American coast: in fact there is a considerable dissimilarity in all these respects. On the other hand there is considerable resemblance in the volcanic nature of the soil, in the climate and size of the islands, between the Galapagos and Cape Verde Archipelagos; but what an absolute difference in their inhabitants! The inhabitants of the Cape Verde Islands are related to those of Africa. Facts such as these admit of no explanation on the ordinary view of independent creation; whereas on the view here maintained it is obvious that the Galapagos Islands would be likely to receive colonists from America and the Cape Verde Islands from Africa.

It is an almost universal rule that the endemic

(*left*) Sea lions and Sally-lightfoot crabs, the Galapagos Islands.

(*right*) A land iguana, one of the two species of iguana found on the Galapagos Islands. Because there are no mammals to compete with them for food, the reptiles tend to grow larger and become the dominant forms of animal life on some islands.

productions of islands are related to those of the nearest continent. The exceptions are few, and most of them can be explained. Thus although Kerguelen Land stands nearer to Africa than to America, the plants are closely related, as we know from Dr. Hooker's account, to those of America: but on the view that this island has been mainly stocked by seeds brought with earth and stones on icebergs drifted by the prevailing currents, this anomaly disappears.

The same law which has determined the relationship between the inhabitants of islands and the nearest mainland is sometimes displayed on a small scale within the same archipelago. Thus each island of the Galapagos Archipelago is tenanted by many distinct species: but these species are related to each other in a very much closer manner than to the inhabitants of America. This is what might have been expected, for islands situated so near to each other would almost necessarily receive immigrants from the same original source and from each other. But how is it that many of the immigrants have been differently modified in islands situated within sight of each other, having the same geological nature, height, climate, etc.?

This difficulty arises in chief part from the deeply seated error of considering the physical conditions of a country as the most important; whereas the nature of the other species with which each has to compete is generally a far more important element of success. Now we find that the species which inhabit the Galapagos Archipelago differ considerably in the several islands. This difference might indeed have been expected if the islands have been stocked by occasional means of transport – a seed, for instance, of one plant having been brought to one island, and that of another plant to another island, though all proceeding from the same general source. Hence, when in former times an immigrant first settled on one island, or when it subsequently spread from one to another, it would undoubtedly be exposed to different conditions in the different islands, for it would have to compete with a different set of organisms. Natural selection would thus favour different varieties in the different islands. Some species, however, might retain the same character throughout the group, just as we see some species spreading throughout a continent and remaining the same.

The really surprising fact is that each new species, after being formed in any one island, did not spread quickly to the other islands. But these islands are separated by deep arms of the sea, in most cases wider than the British Channel; the currents are rapid and sweep between the islands, and gales of wind are extraordinarily rare, so that the islands are far more effectually separated from each other than they appear on a map. Nevertheless some species are common to the several islands; and we may infer from their present distribution that they have spread from one island to the others. But we often take an erroneous view

The finches of the Galapagos present the most striking evidence of immigration by relatively few colonists followed by adaptive radiation into a wide range of vacant ecological niches. All the finches have rather dull plumage and are very similar except in the size and shape of the beaks. Those species that have adapted to seed-eating have large, heavy beaks, those that eat insects have small, sharp beaks, and so on. The finches have evolved into species able to feed on all the types of food resources usually utilized by specialized bird families *(see also the illustration of the woodpecker finch on p. 76).* It is interesting that during territorial fights between males the finches seize each others' beaks, and during courtship the male gives food to the female. In both these instances they are evidently using the beak size to recognize their own kind, in the absence of distinctive plumage differences. This indicates that the species probably evolved from a common ancestor fairly recently; in time, one would expect other, more obvious characteristics to evolve for species identification.

of the probability of closely allied species invading each other's territory. If two species are equally well fitted for their own places, both will probably hold their separate places for almost any length of time. Being familiar with the fact that many species, naturalised through man's agency, have spread with astonishing rapidity over wide areas, we are apt to infer that most species would thus spread; but we should remember that the species which become naturalised in new countries are not generally closely allied to the aboriginal inhabitants, but are very distinct forms.

In the Galapagos Archipelago many of the birds, though so well adapted for flying from island to island, differ on the different islands; thus there are three closely allied species of mocking-thrush, each confined to its own island. Now let us suppose the mocking-thrush of Chatham Island to be blown to Charles Island, which has its own mocking-thrush; why should it succeed in establishing itself there? We may safely infer that Charles Island is well stocked with its own species, for annually more young birds are hatched than can possibly be reared; and the mocking-thrush peculiar to Charles Island is at least as well fitted for its home as is the species peculiar to Chatham Island.

On the same continent, also, preoccupation has probably played a part in checking the commingling of species which inhabit different districts with nearly the same physical conditions. Thus the south-east and south-west corners of Australia have nearly the same physical conditions, yet are inhabited by a vast number of distinct mammals, birds and plants; so it is with the animals inhabiting the great, open and continuous valley of the Amazons.

The same principle which governs the general character of the inhabitants of oceanic islands, namely the relation to the source whence colonists could have been most easily derived, together with their subsequent modification, is of the widest application throughout nature.

Summary of the last and present Chapters
In these chapters I have endeavoured to show that if we make allowance for our ignorance of the effects of changes of climate and of the level of the land, if we remember how ignorant we are with respect to the many means of occasional transport, if we bear in mind how often a species may have ranged continuously over a wide area and then have become extinct in the intermediate tracts, the difficulty is not insuperable in believing that

The Galapagos Islands. Part of the map from *Narrative of the Surveying Voyages of His Majesty's Ships Adventure and Beagle* by Captain Robert Fitzroy.

GALAPAGOS ISLANDS.
By the Officers of
H.M.S.BEAGLE.
1835.

all individuals of the same species, wherever found, are descended from common parents.

As exemplifying the effects of climatal changes on distribution, I have attempted to show how important a part the last Glacial period has played, which, during the alternations of cold in the north and south, allowed the productions of opposite hemispheres to mingle, and left some of them stranded on mountain summits in all parts of the world. As showing how diversified are the means of occasional transport, I have discussed the means of dispersal of fresh-water productions.

If the difficulties be not insuperable in admitting that all the individuals of the same species, and likewise the species of the same genus, have proceeded from one source, then all the leading facts of geographical distribution are explicable on the theory of migration, together with subsequent modification and multiplication of new forms. We can thus understand the high importance of barriers, whether of land or water, in separating the several zoological and botanical provinces. We can understand the concentration of related species within the same areas, and how it is that under different latitudes, for instance in South America, the inhabitants of the plains and mountains, of the forests, marshes and deserts, are linked together in so mysterious a manner, and are likewise linked to the extinct beings which formerly inhabited the same continent. We can see why two areas having nearly the same physical conditions should often be inhabited by very different forms of life, for according to the length of time which has elapsed since the colonists entered the region, according to which forms entered and in greater or lesser numbers, according or not as those which entered came into competition with each other and with the aborigines, and according as the immigrants were capable of varying, there would ensue in the two regions, independent of their physical conditions, infinitely diversified conditions of life: there would be an almost endless amount of organic action and reaction, and we should find some groups of beings greatly, and some only slightly modified, some developed in great force, but others only existing in scanty numbers – and this we do find in the several great geographical provinces of the world.

On these same principles we can understand why oceanic islands should have few inhabitants but that of these a large proportion should be endemic or peculiar, and why, in relation to the means of migration, one group of beings should have all its species peculiar, and another group should have all its species the same with those in an adjoining quarter of the world. We can see why whole groups of organisms, as batrachians [amphibians] and terrestrial mammals, should be absent from oceanic islands, whilst the most isolated islands should possess their own peculiar species of bats. We can see why there should be some relation between the presence of mammals and the depth of the sea between such islands and the mainland. We can clearly see why all the inhabitants of an archipelago, though specifically distinct on the several islets, should be closely related to each other, and should likewise be related, but less closely, to those of the nearest continent.

As the late Edward Forbes often insisted, there is a striking parallelism in the laws of life throughout time and space, the laws governing the succession of forms in past times being nearly the same with those governing, at the present time, the differences in different areas. According to our theory these several relations throughout time and space are intelligible, for whether we look to the allied forms of life which have changed during successive ages, or to those which have changed after having migrated into distant quarters, in both cases they are connected by the same bond of ordinary generation; in both cases the laws of variation have been the same, and modifications have been accumulated by the same means of natural selection.

Chapter 14
Mutual Affinities of Organic Beings:
Morphology: Embryology: Rudimentary Organs

Classification

From the most remote period in the history of the world organic beings have been found to resemble each other in descending degrees, so that they can be classed in groups under groups. This classification is not arbitrary like the grouping of the stars in constellations. The existence of groups would have been of simple significance if one group had been exclusively fitted to inhabit the land, and another the water; one to feed on flesh, another on vegetable matter, and so on; but the case is widely different, for it is notorious how commonly members of even the same subgroup have different habits. Naturalists try to arrange the species, genera, and families in each class on what is called the Natural System. But what is meant by this system? Some authors look at it merely as a scheme for arranging together those living objects which are most alike, and separating those most unlike. Many naturalists think something more is meant by the Natural System; they believe it reveals the plan of the Creator. I believe that community of descent – the one known cause of close similarity in organic beings – is the bond which is partially revealed by our classifications.

Let us consider the rules followed in classification, and the difficulties encountered on the view that classification either gives some unknown plan of creation or is simply a scheme for placing together the forms most like each other. It might have been thought that those parts of the structure which determined the habits of life and the general place of each being in nature would be of high importance in classification. Nothing can be more false. No one regards the external similarity of a whale to a fish as of any importance. Again, no one will say that rudimentary or atrophied organs are of high physiological importance; yet undoubtedly organs in this condition are often of much value in classification.

When naturalists are at work, they do not trouble themselves about the physiological value of the characters which they use in defining a group or allocating any particular species. If they find a character nearly uniform, common to a great number of forms and not common to others, they use it as one of high value; if common to some lesser number, they use it as of subordinate value. If several trifling characters are always found in combination, though no apparent connection can be discovered between them, especial value is set on them.

Characters derived from the embryo are of equal importance with those derived from the adult, for a natural classification includes all ages. But it is not obvious, on the ordinary view, why the embryo should be more important for this purpose than the adult, which alone plays its full part in nature. Yet it has been strongly urged that embryological characters are the most important of all; and this doctrine has very generally been admitted as true.

Classifications are often plainly influenced by chains of affinities. Nothing can be easier than to define a number of characters common to all birds, but with crustaceans any such definition has been found impossible. There are crustaceans at the opposite ends of the series which have hardly a character in common, yet the species at both ends, from being plainly allied to others, and these to others, and so on, can be recognised as belonging to the same class.

Geographical distribution has often been used in classification, more especially in very large groups of closely allied forms. Temminck insists on the utility or even necessity of this practice in certain groups of birds, and it has been followed by several entomologists and botanists.

All the foregoing aids and difficulties in classification may be explained on the view that the Natural System is founded on descent with modification – that the characters which naturalists consider as showing true affinity are those which have been inherited from a common parent, all true classification being genealogical; and that community of descent is the hidden bond which naturalists have been unconsciously seeking.

With species in a state of nature, every naturalist has in fact brought descent into his classification; for he includes in his lowest grade, that of species, the two sexes; and how enormously these sometimes differ in the most important characters is known to every naturalist. The naturalist includes as one species monsters and varieties, not from their partial resemblance to the parent-form, but because they are descended from it. As descent has universally been used in classing together the individuals of the same species, and as it has been used in classing varieties, may not descent have been unconsciously used in grouping species under genera, and genera under higher groups, all under the natural system? I believe it has been unconsciously used; and thus only can I understand the rules followed by our best systematists.

As we have no written pedigrees, we are forced to trace community of descent by resemblances of any kind. Therefore we choose those characters least likely to have been modified by the conditions to which each species has been recently exposed. Rudimentary structures are sometimes even better than other parts of the organisation. We care not how trifling a character may be – if it prevail throughout many species, especially those having very different habits of life, it assumes high value; for we can account for its presence in so many differing forms only by inheritance from a common parent.

We can understand, on the above views, the important distinction between real affinities and analogical or adaptive resemblances. The resemblance in the shape of the body and the fin-like anterior limbs between whales and fishes is analogical. We can clearly understand why analogical or adaptive characters, although of utmost importance to the welfare of the being, are almost valueless to the systematist. For animals belonging to two most distinct lines may have become adapted to similar conditions, and thus have assumed a close external resemblance; but such resemblances will not reveal – will rather tend to conceal – their blood-relationship.

There is a curious class of cases in which close external resemblance does not depend on adaptation to similar habits, but has been gained for protection. I allude to the wonderful manner in which certain butterflies imitate other species. In some districts of South America where, for instance, an *Ithomia* abounds in gaudy swarms, another butterfly, a *Leptalis*, is often found in the same flock; and the latter so closely resembles the *Ithomia* that Mr. Bates, with his eyes sharpened by collecting during eleven years, was continually deceived. The mockers and the mocked are very different in essential structure, and belong not only to distinct genera but often to distinct families.

No less than ten genera include species that imitate other butterflies. The mockers and mocked always inhabit the same region. The mockers are almost invariably rare insects; the mocked in almost every case abound in swarms. In the same district in which a *Leptalis* closely imitates an *Ithomia*, there are sometimes other Lepidoptera mimicking the same *Ithomia*, so that in the same place species of three genera of butterflies and even a moth are found all closely resembling a butterfly

'For animals belonging to two most distinct lines may have become adapted to similar conditions, and thus have assumed a close external resemblance; but such resemblances will not reveal – will rather tend to conceal – their blood-relationship.' **This engraving of a humming-bird hawk moth *(left)* and a humming bird *(right)* is taken from *The Naturalist on the Amazons* by the distinguished nineteenth-century naturalist Henry Bates, who found that they looked so alike when in flight that he often shot a moth by mistake when intending to obtain a specimen of a humming bird. Both animals feed on nectar and they present a striking illustration of convergent evolution.**

belonging to a fourth genus. The imitated form keeps the usual dress of its group, whilst the counterfeiters have changed their dress and do not resemble their nearest allies.

Why has nature condescended to the tricks of the stage? Mr. Bates has hit on the true explanation. The mocked forms, which always abound in numbers, habitually escape destruction to a large extent as they are distasteful to birds and other insect-devouring animals. The mocking forms, on the other hand, are comparatively rare, and belong to rare groups; hence they must suffer habitually from some danger. Now if a member of one of these persecuted groups were to assume a dress like that of a well-protected species, it would often deceive predaceous birds and insects, and thus often escape destruction. Here we have an excellent illustration of natural selection.

Messrs. Wallace and Trimen have likewise described cases of imitation in the Lepidoptera of the Malay Archipelago and Africa, and with some other insects. The much greater frequency of imitation with insects than with other animals is probably the consequence of their small size; insects cannot defend themselves, except the kinds with a sting, and I have never heard of an instance of such kinds mocking other insects, though they are mocked; insects cannot easily escape by flight from larger animals; therefore, speaking metaphorically, they are reduced, like most weak creatures, to trickery and dissimulation.

On the Nature of the Affinities connecting Organic Beings

As the modified descendants of dominant species belonging to the larger genera tend to inherit the advantages which made their groups large and their parents dominant, they are almost sure to spread widely and seize more places in the economy of nature. Thus all organisms are included under a few orders and still fewer classes. As showing how few the higher groups are, and how widely they are spread, the fact is striking that the discovery of Australia has not added an insect belonging to a new class; and that in the vegetable kingdom, as I learn from Dr. Hooker, it has added only two or three families of small size.

In the chapter on Geological Succession I attempted to show, on the principle of each group having diverged much in character during the long-continued process of modification, how it is that the more ancient forms of life often present characters in some degree intermediate between existing groups. As some few of the old forms have transmitted to the present day descendants but little modified, these constitute our aberrant species. The more aberrant any form is, the greater must be the number of connecting forms which have been exterminated. And we have some evidence of aberrant groups having suffered severely from extinction, for they are almost

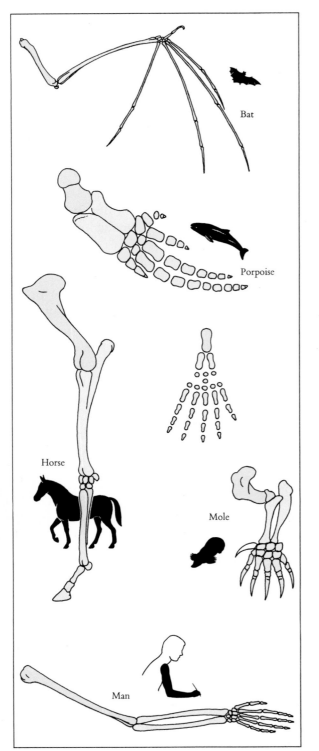

Bat

Porpoise

Horse

Mole

Man

always represented by extremely few species; and such species as do occur are generally very distinct from each other. We can account for this only by looking at aberrant groups as forms which have been conquered by more successful competitors, with a few members still preserved under unusually favourable conditions.

On the principle of the multiplication and gradual divergence of the species descended from a common progenitor, together with their retention by inheritance of some characters in common, we can understand the excessively complex affinities by which all the members of the same family or higher group are connected together. As it is difficult to show the blood-relationship between the numerous kindred of any ancient family even by the aid of a genealogical tree, and almost impossible to do so without this aid, we can understand the extraordinary difficulty which naturalists have experienced in describing, without the aid of a diagram, the various affinities which they perceive between the many living and extinct members of the same natural class.

Extinction, as we have seen in the fourth chapter, has played an important part in defining and widening the intervals between the several groups in each class. We may thus account for the distinct-

On the subject of morphology Darwin writes: '*What can be more curious than that the hand of a man, formed for grasping, that of a mole for digging, the leg of a horse, the paddle of the porpoise, and the wing of the bat should all be constructed on the same pattern, and should include similar bones in the same relative positions?*' The diagram in the centre illustrates the basic plan of the bones found in the forelimbs of all mammals. Any of these bones may have been lengthened, shortened, thickened, diminished or even lost entirely, to adapt the forelimb for different purposes. The most extreme modification shown is that of the horse, where the five sets of long bones that form the hand and fingers in man have become reduced to one set which is thick and strong to take the horse's weight when running. Two fine splints can be seen on either side of the first bone of the set; these splints are derived from two of the other bones, while the two outermost bones have disappeared entirely. In the mole an extra bone has appeared, known as a radial sesamoid, which effectively forms a sixth 'finger'. This bone is derived from a different source altogether, being a development of a bone-like knob within a tendon, and is therefore shown in white.

The placental mammals of Australasia have a unique arrangement in which two of their toe bones are enclosed by the same skin and effectively form one toe. The American opossums, the only marsupials outside Australasia, do not show this feature, suggesting that the marsupials of Australasia were derived from a small and rather unusual stock of ancient marsupials. Shown here are Virginia opossums painted by J. J. Audubon.

ness of whole classes from each other – for instance, of birds from all other vertebrates – by the belief that many ancient forms of life have been utterly lost, through which the early progenitors of birds were connected with the early progenitors of the other and at that time less differentiated vertebrate classes. Extinction has only defined the groups: it has by no means made them; for if every form which ever lived were to reappear, though it would be quite impossible to give definitions by which each group could be distinguished, still a natural classification or arrangement would be possible.

Morphology

We have seen that the members of the same class, independent of their habits of life, resemble each other in the general plan of their organisation. This resemblance is often expressed by the term 'unity of type', or by saying that the several parts in the different species are homologous. What can be more curious than that of the hand of a man,

formed for grasping, that of a mole for digging, the leg of a horse, the paddle of the porpoise, and the wing of the bat should all be constructed on the same pattern, and should include similar bones in the same relative positions?

How curious it is that the hind feet of the kangaroo which are so well fitted for bounding over the open plains, those of the climbing, leaf-eating koala, equally well fitted for grasping the branches of trees, those of the ground-dwelling, insect- or root-eating bandicoots, and those of some other Australian marsupials should all be constructed on the same extraordinary type, namely with the bones of the second and third digits extremely slender and enveloped within the same skin, so that they appear like a single toe with two claws. The case is rendered all the more striking by the American opossums, which follow nearly the same habits of life as some of their Australian relatives, having feet constructed on the ordinary plan. Professor Flower, from whom these statements are taken, remarks in conclusion: 'We may call this conformity to type, without getting much nearer to an explanation of the phenomenon'; and then he adds: 'but it is not powerfully suggestive of true relationship, of inheritance from a common ancestor?'

Geoffroy St. Hilaire has strongly insisted on the high importance of relative position or connexion in homologous parts; they may differ to almost any extent in form and size, and yet remain connected together in the same invariable order. We never find the bones of the arm and forearm or of the thigh and leg transposed. Hence the same names can be given to the homologous bones in widely different animals. In the construction of the mouths of insects, what can be more different than the long spiral proboscis of a sphinx-moth, the folded one of a bee or bug, and the great jaws of a beetle? Yet all these organs, serving for such widely different purposes, are formed by modifications of an upper lip, mandibles, and two pairs of maxillae.

On the theory of the selection of successive slight modifications, the explanation is simple. In changes of this nature there will be little tendency to alter the original pattern or transpose the parts. The bones of a limb might be shortened and flattened to any extent, becoming at the same time enveloped in thick membrane, so as to serve as a fin; or a webbed hand might have its bones lengthened to any extent, with the membrane connecting them increased, so as to serve as a wing; yet these modifications would not alter the framework of the bones or the relative connexion of the parts. If we suppose that an early progenitor of mammals, birds and reptiles had its limbs constructed on the existing pattern, for whatever purpose they served, we can at once perceive the signification of the homologous construction of the limbs throughout the class.

There is another branch of our subject, namely the comparison of the different parts in the same individual. Most physiologists believe that the bones of the skull are homologous – that is, correspond in number and relative connexion – with certain vertebrae. So it is with the wonderfully complex jaws and legs of crustaceans. In a flower the relative position of the sepals, petals, stamens, and pistils, as well as their intimate structure, are intelligible on the view that they consist of metamorphosed leaves arranged in a spire.

How inexplicable are such homologies on the ordinary view of creation! Why should the brain be enclosed in a box of such numerous and extraordinarily shaped pieces of bone, apparently representing vertebrae? Why should one crustacean, which has an extremely complex mouth formed of many parts, consequently always have fewer legs; or conversely, those with many legs have simpler mouths? Why should the sepals, petals, stamens, and pistils in each flower, though fitted for such distinct purposes, be all constructed on the same pattern?

On the theory of natural selection we can, to a certain extent, answer these questions. The un-

known progenitor of the Vertebrata probably possessed many vertebrae; the unknown progenitor of the Articulata, many segments; and the unknown progenitor of flowering plants, many leaves arranged in one or more spires. We have also seen that parts many times repeated are eminently liable to vary, not only in number but in form. Consequently such parts would naturally afford the materials for adaptation to the most different purposes; yet they would generally retain through inheritance traces of their fundamental resemblance.

Development and Embryology
This is one of the most important subjects in the whole round of natural history. The metamorphoses of insects show us what wonderful changes of structure can be effected during development. Such changes, however, reach their acme in the alternate generations of some of the lower animals. Delicate branching corallines studded with polypi and attached to a submarine rock produce, first by budding and then by transverse division, a host of floating jelly-fishes; and then these produce eggs, from which are hatched swimming animalcules which attach themselves to rocks and become developed into branching corallines; and so on in an endless cycle.

Generally the embryos of the most distinct species belonging to the same class are closely similar, but become, when fully developed, widely dissimilar. The larvae of most crustaceans, at corresponding stages of development, closely resemble each other, however different the adults might become; and so it is with very many other animals. In the cat tribe, most of the species when adult are striped or spotted; and stripes or spots can be plainly distinguished in the whelp of the lion and the puma.

The points of structure in which embryos within the same class resemble each other often have no direct relation to their conditions of existence. We cannot suppose that in the embryos of the vertebrata the loop-like arteries near the branchial slits are related to similar conditions in the young mammal nourished in the womb, in the egg of the bird hatched in a nest, and in the spawn of a frog under water. We have no more reason to believe in such a relation than we have to believe that the similar bones in the hand of a man, wing of a bat, and fin of a porpoise are related to similar conditions of life. No one supposes that the stripes on the whelp of a lion, or the spots on the young blackbird, are of any use to these animals. [*Such dappled marks in the young may, in some species, serve as camouflage.*]

The case, however, is different when an animal during any part of its embryonic career is active and has to provide for itself. The adaptation of the larva to its conditions of life is just as perfect as in the adult animal. In most cases, however, the larvae, though active, still obey the law of common embryonic resemblance. Cirripedes afford a good instance of this; even the illustrious Cuvier did not perceive that a barnacle was a crustacean: but a glance at the larva shows this in an unmistakable manner.

The embryo in the course of development generally rises in organisation; in some cases, however, the mature animal must be considered lower in the scale than the larva. To refer again to cirripedes, the larvae in the first stage have loco-motive organs, a simple eye, and a mouth with which they feed largely. In the second stage, answering to the chrysalis of butterflies, they have six pairs of legs, compound eyes, and an-tennae; but they have a closed and imperfect mouth and cannot feed. When their final meta-morphosis is completed they are fixed for life: their legs are now converted into prehensile organs; they again obtain a well-constructed mouth; but they have no antennae, and their eyes are reconverted into a minute simple eye-spot. In this last state cirripedes may be considered as either more highly or more lowly organised than they were in the larval condition. But in some

genera the larvae develop into ordinary herma-phrodites and into complemental males; in the latter the development has assuredly been retro-grade, for the male is a mere sack which lives for a short time and is destitute of every organ of im-portance except those for reproduction.

We are so accustomed to see a difference in structure between embryo and adult that we are tempted to look at this difference as contingent on growth. But there is no reason why, for instance, the wing of a bat or fin of a porpoise should not have been sketched out with all parts in proper proportion as soon as any part became visible. In some whole groups this is the case, and the embryo does not at any period differ widely from the adult: thus there is no metamorphosis in cuttle-fish; the cephalopodic character is manifested long before

A barnacle and a typical crustacean, the shrimp, together with their larvae. '*. . . even the illustrious Cuvier did not perceive that a barnacle was a crustacean: but a glance at the larva shows this in an unmistakable manner.*'

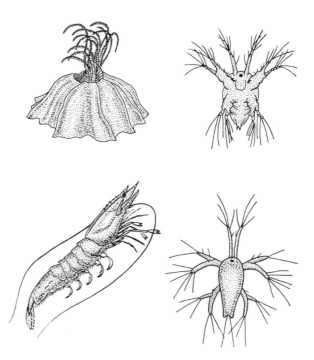

A cougar with young, painted by J. J. Audubon. The spots of the cougar cub, like those of lion cubs, are probably a reversion to the ancestral coat pattern of the cat family, as Darwin suggests. The young of several sorts of mammals and birds have dappled markings, and in some cases these undoubtedly serve as camouflage, but it is unlikely that they do so in young cougars.

the parts of the embryo are completed. Fresh-water crustaceans are born having their proper forms, whilst the marine members of the same class pass through considerable changes during their development. Spiders, again, barely undergo any metamorphosis.

I believe that all these facts can be explained. It is commonly assumed that slight variations necessarily appear in the embryo at an early period. We have little evidence, but what we have certainly points the other way; for it is notorious that breeders of cattle and horses cannot positively tell until some time after birth what will be the merits of their young animals. We see this plainly in our own children; we cannot tell whether a child will be tall or short, or what its precise features will be.

It is of no importance to a very young animal, as long as it remains in its mother's womb or the egg, or as long as it is nourished and protected by its parent, whether most of its characters are acquired a little earlier or later in life. It would not signify, for instance, to a bird which obtained its food by having a much-curved beak whether or not whilst young it possessed a beak of this shape, as long as it was fed by its parents.

I have stated in the first chapter that at whatever age a variation first appears in the parent, it tends to reappear at a corresponding age in the offspring. These two principles, namely that slight variations generally appear at a not very early period of life, and are then inherited at a corresponding, not early, period, explain, as I believe, all the above specified leading facts in embryology.

With some animals successive variations may have supervened at a very early period of life. In this case, the young or embryo will closely re-semble the parent-form; and this is the rule in certain groups, as with cuttle-fish, fresh-water crustaceans, spiders, and some members of the insect class. We can see why the young in such groups do not pass through any metamorphosis from the following contingencies; namely, from the young having to provide at a very early age for their own wants, and from their following the same habits of life with their parents; for in this case it would be indispensable for their existence that they should be modified in the same manner as their parents.

If, on the other hand, it profited the young of an animal to follow slightly different habits from the parent-form, and consequently to be constructed on a slightly different plan, then, on the principle of inheritance at corresponding ages, the young might be rendered by natural selection more and more different from their parents. Differences in the larva might become correlated with successive stages of its development, so that the larva in the first stage might come to differ greatly from the larva in the second stage, as is the case with many animals. The various larval and pupal stages of insects have thus been acquired through adaptation, and not through inheritance from some ancient form.

The case of *Sitaris* – a beetle which passes through certain unusual stages of development – will illustrate how this might occur. The first larval form is an active, minute insect furnished with six legs, two long antennae and four eyes. These larvae are hatched in the nests of bees; and when the male bees emerge from their burrows the larvae spring on them, and afterwards crawl on to the females whilst paired with the males. As soon the female bee deposits her eggs, the larvae of the *Sitaris* leap on the eggs and devour them. Afterwards they undergo a complete change; their eyes disappear; their legs and antennae become rudimentary, and they feed on honey; so they now more closely resemble ordinary insect larvae; ultimately they undergo a further transformation and emerge as the perfect beetle. Now if an insect undergoing transformations like these were to become the progenitor of a new class, the development of the new class would be widely different from that of our existing insects; and the first larval stage certainly would not represent the former condition of any adult and ancient form.

We can now understand how it is that, in the eyes of most naturalists, the structure of the embryo is even more important for classification than that of the adult. In two or more groups of animals, however much they may differ from each other in

Embryos of the dog, bat, rabbit and man, at three stages in their development, as illustrated by Ernst Haeckel in 1891. Only in the final stages do the embryos reveal their identity. Darwin correctly surmises that this is because the mutations that produced the changes between them during their evolution tended to be late-acting. Haeckel, on the other hand, formulated the misleading dogma that every individual must go through the whole evolutionary process of its species during its development, or *'ontogeny recapitulates phylogeny'*.

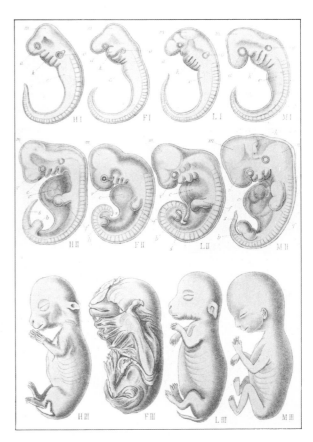

their adult condition, if they pass through closely similar embryonic stages we may feel assured that they all are descended from one parent-form, and are therefore closely related. But dissimilarity in embryonic development does not prove discommunity of descent, for in one or two groups the developmental stages may have been suppressed or so greatly modified through adaptation as to be no longer recognisable.

Rudimentary, Atrophied, and Aborted Organs

Organs or parts in this strange condition, bearing the stamp of inutility, are extremely common throughout nature. It would be impossible to name one of the higher animals in which some part is not in a rudimentary condition. In the mammalia the males possess rudimentary mammae; in snakes one lobe of the lungs is rudimentary; in birds the 'bastard-wing' may safely be considered a rudimentary digit, and in some species the whole wing is so far rudimentary that it cannot be used for flight. What can be more curious than the presence of teeth in foetal whales, which when grown up have not a tooth in their heads; or the teeth, which never cut through the gums, in the upper jaws of unborn calves?

Rudimentary organs in individuals of the same species are liable to vary in the degree of their development and in other respects. In closely allied species the extent to which the same organ has been reduced occasionally differs much. This is well exemplified in the wings of female moths belonging to the same family. In tracing the homologies of any part in the same class, nothing is more common or, in order fully to understand the relations of the parts, more useful than the discovery of rudiments. This is well shown in the drawings by Owen of the leg-bones of the horse, ox and rhinoceros.

Rudimentary organs, such as teeth in the upper jaws of whales and ruminants, can often be detected in the embryo, but afterwards wholly disappear. It is, I believe, a universal rule that a rudimentary part is of greater size in the embryo relative to the adjoining parts than in the adult, so that the organ at this early age is less rudimentary. Hence rudimentary organs in the adult are often said to have retained their embryonic condition.

In reflecting on these facts with respect to rudimentary organs, everyone must be struck with astonishment; for the same reasoning power which tells us that most organs are exquisitely adapted for certain purposes tells us with equal plainness that these rudimentary or atrophied organs are imperfect and useless. In works on natural history rudimentary organs are generally said to have been created 'for the sake of symmetry', or in order 'to complete the scheme of nature'. But this is not an explanation, merely a re-statement of the fact. Nor is it consistent with itself: thus the boa-constrictor has rudiments of hindlimbs and of a pelvis, and if it be said that these bones have been retained 'to complete the scheme of nature', why have they not been retained by other snakes which do not possess even a vestige of these bones? What would be thought of an astronomer who maintained that satellites revolve in elliptic courses round their planets 'for the sake of symmetry', because planets thus revolve round the sun?

On the view of descent with modification, the origin of rudimentary organs is comparatively

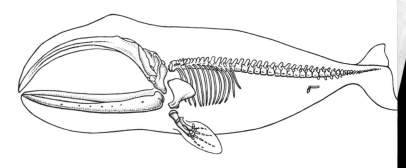

The whales long ago lost their hindlegs when they returned to the sea from the land, but rudimentary bones are found inside a whale which correspond to the pelvic girdle and the upper bone of the hindlimb.

simple. We have plenty of cases of rudimentary organs in our domestic productions – as the stump of a tail in tailless breeds, the reappearance of minute dangling horns in hornless breeds of cattle, more especially in young animals, and the state of the whole flower in the cauliflower. Organs, originally formed by the aid of natural selection, when rendered useless may well be variable, for their variations can no longer be checked by natural selection. All this agrees well with what we see under nature. Moreover, at whatever period of life disuse or selection reduces an organ, and this will generally be when the being has come to maturity and has to exert its full powers, the principle of inheritance at corresponding ages will tend to reproduce the organ in its reduced state at the same mature age, but will seldom affect it in the embryo. Thus we can understand the greater relative size of rudimentary organs in the embryo.

There remains, however, this difficulty. After an organ has ceased being used and become much reduced, how can it be still further reduced in size and finally obliterated? The principle of economy of growth explained in a former chapter, by which the materials forming any part, if not useful to the possessor, are saved as far as is possible, will perhaps come into play in rendering a useless part rudimentary. But this principle will almost necessarily be confined to the earlier stages of the process; for we cannot suppose that a minute organ could be further reduced or absorbed for the sake of economising nutriment.

Finally, as rudimentary organs are the record of a former state of things and have been retained solely through inheritance, we can understand, on the genealogical view of classification, how it is that systematists, in placing organisms in the natural system, have often found rudimentary parts more useful than parts of high physiological importance. Rudimentary organs may be compared with the letters in a word, still retained in the spelling but become useless in the pronunciation, but which serve as a clue for its derivation.

Summary

In this chapter I have attempted to show that the arrangement of all organic beings in groups under groups naturally follows if we admit the common parentage of allied forms, together with their modification through variation and natural selection. On this same view of descent with modification most of the great facts in Morphology become intelligible. On the principle of successive variations not necessarily supervening at a very early period of life, and being inherited at a corresponding period, we can understand the leading facts in Embryology, and on these same principles the occurrence of rudimentary organs might even have been anticipated. The importance of embryological characters and rudimentary organs in classification is intelligible, on the view that a natural arrangement must be genealogical.

The several classes of facts which have been considered in this chapter seem to me to proclaim so plainly that the innumerable species, genera and families with which this world is peopled are all descended, each within its own class or group, from common parents, and have all been modified in the course of descent, that I should without hesitation adopt this view, even if it were unsupported by other facts or arguments.

Chapter 15

Conclusion

I have now given the facts and considerations which have thoroughly convinced me that species have been modified during a long course of descent. This has been effected chiefly through the natural selection of numerous successive, slight, favourable variations; aided in an important manner by the inherited effects of the use and disuse of parts; and in an unimportant manner, that is in relation to adaptive structures, whether past or present, by the direct action of external conditions and by variations which seem to us in our ignorance to arise spontaneously. It appears that I formerly underrated the frequency and value of these latter forms of variation, as leading to permanent modifications of structure independently of natural selection. But as my conclusions have lately been much misrepresented, and it has been stated that I attribute the modification of species exclusively to natural selection, I may be permitted to remark that in the first edition of this work, and subsequently, I placed in a most conspicuous position – namely, at the close of the Introduction – the following words: 'I am convinced that natural selection has been the main but not the exclusive means of modification.' This has been of no avail. Great is the power of steady misrepresentation; but the history of science shows that fortunately this power does not long endure.

It can hardly be supposed that a false theory would explain in so satisfactory a manner as does the theory of natural selection the several large classes of facts above specified. It has recently been objected that this is an unsafe method of arguing; but it is a method used in judging of the common events of life, and has often been used by the greatest natural philosophers. The undulatory theory of light has thus been arrived at; and the belief in the revolution of the earth on its own axis was until lately supported by hardly any direct evidence. It is no valid objection that science as yet throws no light on the far higher problem of the essence or origin of life. Who can explain what is the essence of the attraction of gravity? No one now objects to following out the results consequent on this unknown element of attraction; notwithstanding that Leibnitz formerly accused Newton of introducing 'occult qualities and miracles into philosophy.'

I see no good reason why the views given in this volume should shock the religious feelings of anyone. It is satisfactory, as showing how transient such impressions are, to remember that the greatest discovery ever made by man, namely the law of the attraction of gravity, was also attacked by Leibnitz, 'as subversive of natural, and inferentially of revealed, religion.' A celebrated author and divine has written to me that 'he has gradually learnt to see that it is just as noble a conception of the Deity to believe that He created a few original forms capable of self-development into other and needful forms, as to believe that He required a fresh act of creation to supply the voids caused by the action of His laws.'

Why, it may be asked, until recently did nearly all the most eminent living naturalists and geologists disbelieve in the mutability of species? It cannot be asserted that organic beings in a state of nature are subject to no variation; it cannot be proved that the amount of variation in the course of long ages is a limited quantity; no clear distinction has been, or can be, drawn between species and well-marked varieties. It cannot be maintained that species when intercrossed are invariably sterile and varieties invariably fertile; or that sterility is a special endowment and sign of creation. The belief that species were immutable productions was almost unavoidable as long as the history of the world was thought to be of short duration; and now that we have acquired some idea of the lapse of time, we are too apt to assume, without proof, that the geological record is so perfect that it would have afforded us plain evidence of the mutation of species if they had undergone mutation.

But the chief cause of natural unwillingness to admit that one species has given birth to other and distinct species is that we are always slow in admitting great changes of which we do not see the steps. The difficulty is the same as that felt by so many geologists when Lyell first insisted that long lines of inland cliffs had been formed, and great valleys excavated, by the agencies which we see still at work. The mind cannot possibly grasp the full meaning of the term of even a million years; it cannot add up and perceive the full effects of many slight variations accumulated during an almost infinite number of generations.

Although I am fully convinced of the truth of the views given in this volume, I by no means expect to convince experienced naturalists whose minds are stocked with a multitude of facts all viewed, during a long course of years, from a point of view directly opposite to mine. It is so easy to hide our ignorance under such expressions as the 'plan of creation,' 'unity of design,' etc., and to think that we give an explanation when we only re-state a fact. Anyone whose disposition leads him to attach more weight to unexplained difficulties than to the explanation of a certain number of facts will certainly reject the theory. A few naturalists, endowed with much flexibility of mind, and who have already begun to doubt the immutability of species, may be influenced by this

'I see no good reason why the views given in this volume should shock the religious feelings of anyone.' Nevertheless they did, particularly the idea that man was descended from an ape-like stock, which, though Darwin did not state it plainly, was implicit in *The Origin of Species*. Bishop Wilberforce of Oxford, Darwin's most implacable enemy among the clergy, wrote that Christianity 'was utterly irreconcilable with the degrading notion of the brute origin of him who was created in the image of God.' The wife of the Bishop of Worcester put it rather more candidly when she said: *'Descended from the apes! My dear, let us hope that it is not true, but if it is, let us pray that it will not become generally known.'*

volume; but I look with confidence to the future, to young and rising naturalists who will be able to view both sides of the question with impartiality. Whoever is led to believe that species are mutable will do good service by conscientiously expressing his conviction; for thus only can the load of prejudice by which this subject is overwhelmed be removed.

Several eminent naturalists have of late published their belief that a multitude of reputed species in each genus are not real species, but that other species are real, that is, have been independently created. This seems to me a strange conclusion to arrive at. They admit that a multitude of forms, which till lately they themselves thought were special creations, and which have all the characteristic features of true species, have been produced by variation, but they refuse to extend the same view to other and slightly different forms. Nevertheless they do not pretend that they can define, or even conjecture, which are the created forms of life and which are those produced by secondary laws. The admit variation as a *vera causa* in one case, they arbitrarily reject it in another, without assigning any distinction in the two cases. The day will come when this will be given as a curious illustration of the blindness of preconceived opinion.

These authors seem no more startled at a miraculous act of creation than at an ordinary birth. But do they really believe that at innumerable periods in the earth's history certain elemental atoms have been commanded suddenly to flash into living tissues? Do they believe that at each supposed act of creation one individual or many were produced? Were all the infinitely numerous kinds of animals and plants created as eggs or seed, or as full grown? and in the case of mammals, were they created bearing the false marks of nourishment from the mother's womb? Undoubtedly some of these same questions cannot be answered by those who believe in the appearance or creation of only a few forms of life, or of some

one form alone. It has been maintained by several authors that it is as easy to believe in the creation of a million beings as of one; but Maupertuis's philosophical axiom 'of least action' leads the mind more willingly to admit the smaller number; and certainly we ought not to believe that innumerable beings within each great class have been created with plain but deceptive marks of descent from a single parent.

As a record of a former state of things, I have retained in the foregoing paragraphs, and elsewhere, several sentences which imply that naturalists believe in the separate creation of each species, and I have been much censured for having thus expressed myself. But undoubtedly this was the general belief when the first edition of the present work appeared. I formerly spoke to very many naturalists on the subject of evolution, and never once met with any sympathetic agreement. It is probable that some did then believe in evolution, but they were either silent, or expressed themselves so ambiguously that it was not easy to understand their meaning. Now things are wholly changed and almost every naturalist admits the great principle of evolution. There are, however, some who still think that species have suddenly given birth, through quite unexplained means, to new and totally different forms: but, as I have attempted to show, weighty evidence can be opposed to the admission of great and abrupt modifications. Under a scientific point of view little advantage is gained by believing that new forms are suddenly developed in an inexplicable manner from different forms, over the old belief in the creation of species from the dust of the earth.

It may be asked how far I extend the doctrine of the modification of species. The question is difficult to answer, because the more distinct the forms are which we consider, by so much the arguments in favour of community of descent become fewer in number and less in force. But some arguments of the greatest weight extend very far. All the members of whole classes are connected together by a chain of affinities, and all can be classed on the same principle in groups subordinate to groups. Fossil remains sometimes tend to fill up very wide intervals between existing orders.

Organs in a rudimentary condition plainly show that an early progenitor had the organ in a fully developed condition; and this in some cases implies an enormous amount of modification in the descendants. Throughout whole classes various structures are formed on the same pattern, and at a very early age the embryos closely resemble each other. Therefore I cannot doubt that the theory of descent with modification embraces all the members of the same great class or kingdom. I believe that animals are descended from at most four or five progenitors, and plants from an equal or lesser number.

Analogy would lead me one step farther, namely to the belief that all animals and plants are descended from some one prototype. But analogy may be a deceitful guide. Nevertheless all living things have much in common, in their chemical composition, their cellular structure, their laws of growth, and their liability to injurious influences. We see this even in so trifling a fact as that the same poison often similarly affects plants and animals; or that the poison secreted by the gall-fly produces monstrous growths on the wild rose or oak-tree. With all organic beings, except perhaps some of the very lowest, sexual reproduction seems to be essentially similar.

If we look even to the two main divisions – namely, to the animal and vegetable kingdoms – certain low forms are so far intermediate in character that naturalists have disputed to which kingdom they should be referred. As Professor Asa Gray has remarked, 'the spores and other reproductive bodies of many of the lower algae may claim to have first a characteristically animal, and then an unequivocally vegetable existence.' Therefore, on the principle of natural selection with divergence of character, it does not seem

incredible that, from some such low and intermediate form, both animals and plants may have been developed; and if we admit this we must likewise admit that all the organic beings which have ever lived on this earth may be descended from some one primordial form.

But this inference is chiefly grounded on analogy, and it is immaterial whether or not it be accepted. No doubt it is possible that at the first commencement of life many different forms were evolved; but if so, we may conclude that only a very few have left modified descendants. For, as I have remarked in regard to each great kingdom, we have distinct evidence in their embryological, homologous and rudimentary structures that all the members are descended from a single progenitor.

When the views advanced by me in this volume and by Mr. Wallace, or when analogous views on the origin of species are generally admitted, we can dimly foresee that there will be a considerable revolution in natural history. Systematists will be able to pursue their labours as at present; but they will not be incessantly haunted by the shadowy doubt whether this or that form be a true species. This, I feel sure and I speak after experience, will be no slight relief. The endless disputes whether or not some fifty species of British brambles are good species will cease. Systematists will have only to decide (not that this will be easy) whether any form be sufficiently constant and distinct from other forms to be capable of definition; and if definable, whether the differences be sufficiently important to deserve a specific name.

Hereafter we shall be compelled to acknowledge that the only distinction between species and well-marked varieties is that the latter are known, or believed, to be connected at the present day by intermediate gradations, whereas species were formerly thus connected. Hence, without rejecting the consideration of the present existence of intermediate gradations between any two forms,

'*Analogy would lead me one step farther, namely to the belief that all animals and plants are descended from some one prototype. But analogy may be a deceitful guide. Nevertheless all living things have much in common, in their chemical composition, their cellular structure, their laws of growth, and their liability to injurious influences.*' It is now established that all living things do share a common ancestor. The picture shows microscopic organisms known as *Euglena* which have a feature characteristic of animals (they are able to move around), and another characteristic of plants (they are green and can photosynthesize, that is make their own food with the aid of sunlight). Such organisms are related to the ancestral stock of single-celled organisms from which both multicellular plants and multicellular animals developed. The study of biochemistry has provided even more telling evidence for the unity of life, since it has revealed that the basic metabolic processes are identical in plants, animals and micro-organisms, including the most primitive living beings, bacteria.

natural history will rise greatly in interest. The terms used by naturalists of affinity, relationship, community of type, paternity, morphology, adaptive characters, rudimentary and aborted organs, etc., will cease to be metaphorical and will have a plain signification. When we no longer look at an organic being as a savage looks at a ship, as something wholly beyond his comprehension; when we regard every production of nature as one which has had a long history; when we contemplate every complex structure and instinct as the summing up of many contrivances, each useful to the possessor, in the same way as any great mechanical invention is the summing up of the labour, the experience, the reason, and even the blunders of numerous workmen; when we thus view each organic being, how far more interesting – I speak from experience – does the study of natural history become!

A grand and almost untrodden field of inquiry will be opened on the causes and laws of variation, on correlation, on the effects of use and disuse, on the direct action of external conditions, and so forth. The study of domestic productions will rise immensely in value. A new variety raised by man will be a more important and interesting subject for study than one more species added to the infinitude of already recorded species. Our classifications will come to be, as far as they can be, genealogies, and will then truly give what may be called the plan of creation. The rules for classifying will no doubt become simpler when

we shall be led to weigh more carefully and to value higher the actual amount of difference between them. It is quite possible that forms now generally acknowledged to be merely varieties may hereafter be thought worthy of specific names, and in this case scientific and common language will come into accordance. In short, we shall have to treat species in the same manner as those naturalists treat genera who admit that genera are merely artificial combinations made for convenience. This may not be a cheering prospect; but we shall at least be freed from the vain search for the undiscovered and undiscoverable essence of the term species.

The other and more general departments of

'. . . *when we regard every production of nature as one which has had a long history: when we contemplate every complex structure and instinct as the summing up of many contrivances, each useful to the possessor . . . how far more interesting – I speak from experience – does the study of natural history become!*' Both Darwin and Wallace remained practical naturalists, but with their views of the living world informed and inspired by evolutionary ideas. While Darwin conducted his careful experiments on plants, dissected barnacles and studied earthworms, Wallace carried out pioneering work on the distribution of animals. He identified various zoogeographical regions, as described in his major work *The Geographical Distribution of Animals* from which this engraving of animals in Borneo is taken. Notice the flying lemurs which Darwin discusses on p 106.

we have a definite object in view. We possess no pedigrees or armorial bearings; we have to trace the lines of descent in our natural genealogies by characters of any kind which have long been inherited. Rudimentary organs will speak infallibly with respect to the nature of long-lost structures. Species and groups of species which are called aberrant, and which may fancifully be called living fossils, will aid us in forming a picture of the ancient forms of life. Embryology will often reveal to us the structure, in some degree obscured, of the prototypes of each great class.

When we can feel assured that all the individuals of the same species, and all the closely allied species of most genera, have within a not very remote period descended from one parent and migrated from some one birth-place; and when we better know the many means of migration, then, by the light which geology throws on former changes of climate and of the level of the land, we shall surely be enabled to trace the former migrations of the inhabitants of the whole world. Even at present, by comparing the differences between the inhabitants of the sea on the opposite sides of a continent, and the nature of the various inhabitants on that continent in relation to their apparent means of immigration, some light can be thrown on ancient geography.

The noble science of Geology loses glory from the extreme imperfection of the record. The crust of the earth with its embedded remains must not be looked at as a well-filled museum but as a poor collection made at hazard and at rare intervals. The accumulation of each great fossiliferous formation will be recognised as having depended on an unusual concurrence of favourable circumstances, and the blank intervals between the successive stages as having been of vast duration. We shall be able to gauge with some security the duration of these intervals by a comparison of the preceding and succeeding organic forms. But we must be cautious in attempting to correlate as strictly contemporaneous two formations, which

do not include many identical species, by the general succession of the forms of life. As species are produced and exterminated by slowly acting and still existing causes, and as the most important of all causes of organic change is one which is almost independent of altered physical conditions, namely the mutual relation of organism to organism, the improvement of one organism entailing the improvement or the extermination of others, it follows that the amount of organic change in the fossils of consecutive formations probably serves as a fair measure of the relative, though not actual, lapse of time. A number of species, however, keeping in a body might remain for a long period unchanged, whilst within the same period several of these species, by migrating into new countries and coming into competition with foreign associates, might become modified; so that we must not overrate the accuracy of organic change as a measure of time.

In the future I see open fields for far more important researches. Psychology will be securely based on the foundation already well laid by Mr. Herbert Spencer, that of the necessary acquirement of each mental power and capacity by gradation. Much light will be thrown on the origin of man and his history.

Authors of the highest eminence seem to be fully satisfied with the view that each species has been independently created. To my mind it accords better with what we know of the laws impressed on matter by the Creator that the production and extinction of the past and present inhabitants of the world should have been due to secondary causes like those determining the birth and death of the individual. When I view all beings not as special creations but as the lineal descendants of some few beings which lived long before the first bed of the Cambrian system was deposited, they seem to me to become ennobled.

Judging from the past, we may safely infer that not one living species will transmit its unaltered likeness to a distant futurity. And of the species

'. . . we may feel certain that the ordinary succession by generation has never once been broken, and that no cataclysm has desolated the whole world.' **The flood described by the Bible was popularly invoked to reconcile the nineteenth-century discoveries of fossils of extinct animals with orthodox religious beliefs about the Creation. For this reason dinosaurs and other extinct beings were often referred to as 'ante-diluvian monsters'. This engraving shows mammoths vainly trying to escape the rising flood waters in Asia.**

now living very few will transmit progeny of any kind to a far distant futurity; for the manner in which all organic beings are grouped shows that the greater number of species in each genus, and all the species in many genera, have left no descendants, but have become utterly extinct. We can so far take a prophetic glance into futurity as to foretell that it will be the common and widely spread species, belonging to the larger and dominant groups within each class, which will ultimately prevail and procreate new and dominant species.

As all the living forms of life are the lineal descendants of those which lived long before the Cambrian epoch, we may feel certain that the ordinary succession by generation has never once been broken, and that no cataclysm has desolated the whole world. Hence we may look with some confidence to a secure future of great length. And as natural selection works solely by and for the good of each being, all corporeal and mental endowments will tend to progress towards perfection.

It is interesting to contemplate a tangled bank, clothed with many plants of many kinds, with birds singing on the bushes, with various insects flitting about, and with worms crawling through the damp earth, and to reflect that these elaborately constructed forms, so different from each other and dependent upon each other in so complex a manner, have all been produced by laws acting around us. These laws, taken in the largest sense, being Growth with Reproduction; Inheritance, which is almost implied by reproduction; Variability from the indirect and direct action of the conditions of life, and from use and disuse; a Ratio of Increase so high as to lead to a Struggle for Life, and as a consequence to Natural Selection, entailing Divergence of Character and the Extinction of less-improved forms. Thus, from the war of nature, from famine and death, the most exalted object which we are capable of conceiving, namely the production of the higher animals, directly follows. There is grandeur in this view of life, with its several powers, having been originally breathed by the Creator into a few forms or into one; and that, whilst this planet has gone cycling on according to the fixed law of gravity, from so simple a beginning endless forms most beautiful and most wonderful have been and are being evolved.

Glossary

The glossary to the original work has been abridged, and the definitions revised in some cases. Definitions have also been added to cover the Introduction. All major additions are shown in *italic*. For the names of geological strata the reader is referred to the inside cover of the book.

Aberrant Forms or groups of animals or plants which deviate in important characters from their nearest allies.

Aborted An organ is said to be aborted when its development has been arrested at a very early stage.

Albinism Albinos are animals in which the usual colouring matters have not been produced in the skin and its appendages. Albinism is the state of being an albino.

Algae A class of plants including the seaweeds and the filamentous fresh-water weeds.

Allometry, Allometric growth *An organism is said to show allometric growth if the rates of growth of separate parts differ, so that these parts do not bear the same proportions to each other in individuals of different sizes. Allometry can be observed between individuals of the same species, and between related species, e.g. in the size of the antlers relative to the body in various species of deer.*

Alternation of generations This term is applied to a peculiar mode of reproduction which prevails among many of the lower animals, in which the egg produces a living form quite different from its parent, but from which the parent form is reproduced.

Ammonites A group of fossil, spiral, chambered shells, allied to the existing pearly nautilus, but having the partitions between the chambers waved in complicated patterns at their junction with the outer wall of the shell.

Amphibians *See Batrachians.*

Amphioxus *See Lancelet.*

Analogy That resemblance of structures which depends upon similarity of function, as in the wings of insects and birds. Such structures are said to be analogous, and to be analogues of each other.

Animalcule A minute animal: generally applied to those visible only by the microscope. (*This term is no longer used.*)

Annelids A class of worms which includes the ordinary marine worms, the earthworms, and the leeches.

Antennae Jointed organs appended to the head in insects, crustacea and centipedes.

Anthers The summits of the stamens of flowers, in which the pollen is produced.

Aplacentalia or **Aplacental mammals** See Mammalia.

Archetypal Of, or belonging to, the archetype, or ideal primitive form upon which all the beings of a group seem to be organised.

Articulata (*now known as Arthropoda*) A great division of the animal kingdom characterised generally by having the surface of the body divided into rings called segments, a greater or less number of which are furnished with jointed legs; (includes: insects, crustaceans and centipedes).

Atrophied Arrested in development at a very early stage.

Arthropoda *See Articulata.*

Balanus The genus including the common barnacles which live in abundance on the rocks of the sea-coast.

Batrachians (*now known as Amphibians*) A class of animals undergoing a peculiar metamorphosis, in which the young animal is generally aquatic and breathes by gills; (includes: frogs, toads, and newts).

Brachiopoda A class of marine molluscs furnished with a bivalve shell, attached to submarine objects by a stalk and furnished with fringed arms, by the action of which food is carried to the mouth.

Branchiae Gills; organs for respiration in water.

Canidae The dog family, including the dog, wolf, fox and jackal.

Carapace The shell enveloping the anterior part of the body in crustaceans generally; applied also to the hard, shelly pieces of the cirripedes.

Carbonic acid *A very weak acid formed in rainwater by carbon dioxide gas (which occurs naturally in the atmosphere) becoming dissolved in the rain as it falls.*

Caudal Of, or belonging to, the tail.

Cell *Living bodies are made up of units of microscopic size, known as cells. These are separated from adjoining cells by membranes.*

Cephalopods The highest class of the molluscs, characterised by having the mouth surrounded by fleshy arms or tentacles, which in most living species are furnished with sucking-cups; (examples: cuttle-fish, octopus, nautilus).

Cetacea An order of mammals, including the whales and dolphins, having the form of the body fish-like, the skin naked, and only the forelimbs developed.

Chelonia An order of reptiles including the turtles and tortoises.

Chromosomes *Strands which can be observed with the aid of a microscope in the nuclei of all eukaryotic cells. They carry the genes, which are composed of DNA.*

Chrysalis *See Pupa.*

Cirripedes An order of crustaceans including the barnacles. Their young resemble those of many other crustaceans in form, but when mature they are always attached to other objects, either by means of a stalk, or directly, and their bodies are enclosed by a calcareous shell composed of several pieces, two of which can open to give issue to a bunch of curled, jointed tentacles, which represent the limbs.

Cocoon A case, usually of silky material, in which insects are frequently enveloped during the second or resting stage (pupa) of their existence.

Coleoptera Beetles; an order of insects having a biting mouth, and the first pair of wings *(elytra)* more or less horny, forming sheaths for the second pair, and usually meeting in a straight line down the middle of the back.

Column A peculiar organ in the flowers of orchids, in which the stamens, style and stigma (or the reproductive parts) are united.

Compositae Plants in which the inflorescence consists of numerous small flowers (florets) in a dense head, the base of which is enclosed by a common envelope; (examples: the daisy and dandelion).

Convergent evolution *If two forms are descended from very different ancestors but show a superficial similarity to each other through adaptation to similar niches, they are said to show convergent evolution; e.g. whales and fishes.*

Corolla *A collective term for the petals of a flower, used especially when they are fused into a tube for part or all of their length.*

Cotyledons The first leaves, or seed leaves, of plants.

Crossing over *The exchange of parts of homologous chromosomes during meiosis; see diagram on p. 24.*

Crustaceans A class of articulated animals, having the skin of the body generally more or less hardened by the deposition of calcareous matter and breathing by means of gills; (examples: crab, lobster and shrimp).

Cutaneous Of, or belonging to, the skin.

Degradation The wearing down of the land by the action of the sea or of meteoric agents.

Denudation The wearing away of the surface of the land by water.

Dicotyledons or **Dicotyledonous plants** A class of plants characterised by having two seed leaves, and by the reticulation of the veins of leaves.

Differentiation The separation or discrimination of parts or organs which in simpler forms of life are more or less united.

Dimorphic Having two distinct forms.

Dioecious Having the organs of the sexes upon distinct individuals.

Dominant *Of a pair of different genes for the same character the one which will be expressed if the two different genes are present. A gene may be dominant in the presence of certain different genes but not others, or its dominance may depend on the environment.*

Dorsal Of, or belonging to, the back.

Ecology *The study of the interrelations of living organisms with each other and with their environment.*

Ecosystem *A term used for the totality of ecological interactions within a physically defined space. It includes all the living organisms and all such non-living substances as interact with those living organisms, in an area such as a pond, a wood or a small island.*

Edentata A peculiar order of mammals, characterised by the absence of at least the middle incisor (front) teeth in both jaws; (examples: the sloths and armadillos).

Elytra The hardened forewings of beetles, serving as sheaths for the membranous hindwings, which constitute the true organs of flight.

Embryo The young animal undergoing development within the egg or womb.

Embryology The study of the development of the embryo.

Endemic Peculiar to a given locality.

Enzymes *Proteins which act as catalysts in chemical reactions taking place in the body. Hundreds of different enzymes are found in every living organism. By determining* *which reactions occur, and at what rate, enzymes control the complex chemical transformations necessary for life.*

Ephemerous insects Insects allied to the may-fly.

Eukaryote *An organism which has its chromosomes contained within a nucleus, and which is distinguished from the simpler prokaryotes by other cellular structures and certain differences in biochemical function. The eukaryotes include higher plants and animals, protozoa, fungi, and algae other than blue-green algae.*

Fauna The totality of the animals naturally inhabiting a certain country or region, or which have lived during a given geological period.

Felidae The cat family.

Feral Having become wild from a state of cultivation or domestication.

Fitness *The fitness of an organism is defined as its ability to pass on its genes to subsequent generations, relative to the ability of other individuals of the same species to do so.*

Flora The totality of the plants growing naturally in a country or during a given geological period.

Florets Flowers imperfectly developed in some respects, and collected into a dense spike or head, as in the grasses and in members of the Compositae, such as the dandelion.

Foetal Of, or belonging to, the foetus, or embryo in course of development.

Foraminifera A class of animals of very low organisation, and generally of small size, having a jelly-like body, from the surface of which delicate filaments can be given off and retracted for the prehension of external objects, and having a calcareous or sandy shell.

Fossorial Having the faculty of digging. The fossorial hymenoptera are a group of wasp-like insects which burrow in sandy soil to make nests for their young.

Fungi (sing. **Fungus**) A class of cellular plants of which mushrooms, toadstools, and moulds are familiar examples.

Gallinaceous birds An order of birds of which the common fowl, turkey and pheasant are well-known examples.

Gamete *A cell having half the usual number of chromosomes, which unites with another gamete during sexual reproduction to give a fertilized egg. In animals male gametes are known as sperm, female gametes as eggs or ova.*

Ganglion A swelling or knot from which nerves are given off as from a centre.

Ganoid fishes Fishes covered with peculiar enamelled bony scales. Most of them are extinct.

Gene *The name now given to the Mendelian unit of inherit-*

ance. In genetic terms it is the smallest, indivisible unit of heredity; in biochemical terms it is a length of DNA which codes for a single protein.

Gene recombination The process whereby any individual receives an entirely unique combination of the genes possessed by its parents. Gene recombination takes place during meiosis when each gamete receives only one of each pair of homologous chromosomes possessed by the parent, and when parts of homologous chromosomes are exchanged through crossing over.

Genetic drift *The random change in gene frequencies due to chance. It is most evident in small, isolated communities.*

Genetic fitness *See Fitness.*

Genotype *The total genetic information carried by an organism.*

Gills Organs for respiration in water.

Glacial period A period of great cold and of enormous extension of ice upon the surface of the earth. It is believed that glacial periods have occurred repeatedly during the geological history of the earth, but the term is generally applied to the *Pleistocene period* when nearly the whole of Europe was subjected to an arctic climate.

Glottis The opening of the windpipe into the oesophagus or gullet.

Gneiss A rock approaching granite in composition but really produced by the alteration of a sedimentary deposit after its consolidation. *(See Metamorphic rocks.)*

Granite A *(plutonic)* rock consisting essentially of crystals of felspar and mica in a mass of quartz.

Habitat The locality in which a plant or animal naturally lives.

Hermaphrodite Possessing the organs of both sexes.

Heterozygous *An organism is said to be heterozygous for the genes for a certain character if the two genes it carries for that character are not identical.*

Homology That relation between parts which results from their development from corresponding embryonic parts, either in different animals, as in the case of the arm of man and the wing of a bird, or in the same individual, as in the case of the forelegs and hindlegs of mammals, and the segments of which the body of a worm is composed. The latter is called serial homology. The parts which stand in such a relation to each other are said to be homologous, and one such part or organ is called the homologue.

Homologous chromosomes *Chromosomes which pair up with each other during meiosis; the homologue of a chromosome is sometimes referred to as its 'sister chromosome'.*

Homozygous *An organism is said to be homozygous for the genes for a certain character if the two genes it carries for that character are identical.*

Hybrid The offspring of the union of two distinct species.

Hymenoptera An order of insects possessing biting jaws and usually four membranous wings in which there are a few veins; (includes: bees, wasps and ants).

Hypertrophied Excessively developed.

Ichneumonidae A family of hymenopterous insects, the members of which lay their eggs in the bodies or eggs of other insects.

Indigenes The aboriginal animal or vegetable inhabitants of a country or region.

Industrial melanism *The occurrence, in areas with air pollution, of blackish varieties of species which are normally a lighter colour.*

Inflorescence The mode of arrangement of the flowers of plants. (*A term used for the whole flower branch of a plant, including flowers, flower stalks and bracts.*)

Infusoria A class of microscopic animalcules, so called from their having originally been observed in infusions of vegetable matters. They consist of a gelatinous material enclosed in a delicate membrane, the whole or part of which is furnished with short vibrating hairs (called cilia) by means of which the animalcules swim through the water or convey food to the mouth. (*Pasteur's demonstration that spontaneous generation does not occur made this term obsolete; see Protozoa.*)

Insectivora *An order of mammals including the shrews, moles and hedgehogs.*

Insectivorous Feeding on insects.

Invertebrata or **Invertebrate animals** Those animals which do not possess a backbone or spinal (*vertebral*) column.

Lacustrine deposits *Deposits of sediment in lakes.*

Lancelet *A marine animal which inhabits shallow water and externally resembles a small fish. It belongs to a group known as the Cephalochordates, which may be ancestral to the vertebrates. It has a stiffening rod, known as a notochord, rather than a vertebral column, and it lacks a skull.*

Larva (pl. **Larvae**) The first condition of an insect at its issuing from the egg, when it is usually in the form of a caterpillar or maggot.

Larynx The upper part of the windpipe opening into the gullet.

Laurentian A group of greatly altered and very ancient (*Precambrian*) rocks which is greatly developed along the course of the St. Lawrence, in Canada.

Leguminosae An order of plants represented by the common peas and beans.

Lemuridae or **Lemurs** A group of four-handed mammals, distinct from the monkeys and approaching the Insectivores in some of their characters and habits. Its members have the nostrils curved or twisted, and a claw instead of a nail upon the first finger of the hind hands. (*Note that the 'flying lemur' is not a lemur at all, but occupies an order of its own, most closely related to the Insectivores.*)

Lepidoptera An order of insects which includes the butterflies and moths.

Linked genes *Genes which occur on the same chromosome are said to be linked, because they tend to be inherited together. The closer the genes are to each other the less likely it is that they will be separated by crossing over (q.v.), and the probability of their being inherited together increases. Genes which occur very near to each other are said to be closely or tightly linked.*

Littoral Inhabiting the sea shore.

Mammalia The highest class of animals, including the ordinary hairy quadrupeds, the whales and man, and characterized by the production of living young which are nourished after birth by milk from the teats (mammary glands) of the mother. A striking difference in the embryonic development has led to the division of this class into two great groups; in one of these a vascular connection, called the placenta, is formed between the embryo and the mother; in the other this is wanting, and the young are produced in a very incomplete state. The former including the greater part of the class, are called placental mammals; the latter or aplacental mammals include the marsupials and monotremes.

Mandibles In insects the first or uppermost pair of jaws, which are generally solid, horny, biting organs. In birds the term is applied to both jaws with their horny coverings.

Marsupials An order of mammals in which the young are born in a very incomplete state of development, and carried by the mother, while sucking, in a ventral pouch called the marsupium; (examples: kangaroos and opossums).

Meiosis *The process of cell division by which gametes are formed. There are two separate divisions during meiosis, but the chromosomes replicate only once so that the gametes have only half the usual number of chromosomes.*

Melanism The opposite of albinism; an undue development of colouring material in the skin and its appendages. (*See also Industrial melanism.*)

Metamorphic rocks Sedimentary rocks which have undergone alteration, generally by the action of heat, subsequent to their deposition and consolidation.

Microfossils *Fossils of organisms of microscopic size.*

Micropalaeontology *The study of microfossils.*

Mitosis *The normal process of cell division in which two identical daughter cells are produced.*

Mollusca One of the great divisions of the animal kingdom, including those animals which have a soft body, usually furnished with a shell; the cuttle-fish, common snails, whelks, oysters, mussels, and cockles may serve as examples of them.

Mongrel *The offspring of a cross between two distinct varieties of the same species.*

Monocotyledons or **Monocotyledonous plants** Plants in which the seed sends up only a single seed leaf (or cotyledon); characterized by the veins of the leaves being generally straight; (examples: grasses, lilies, orchids and palms).

Monotremes *A group of mammals which show several primitive features, notably laying eggs rather than producing their young alive. They are found only in Australia and New Guinea and the sole representatives of the group are the duck-billed platypus (Ornithorhynchus) and five species of spiny anteaters (Echidna).*

Moraines The accumulations of fragments of rock brought down by glaciers.

Morphology The law of form or structure independent of function.

Mutation *A spontaneous change in a gene. This can result from inaccurate copying of the DNA during replication, so that a base is lost or added, or one base is substituted for another, or from the loss or inversion of a segment of DNA.*

Nascent Commencing development.

Neuration The arrangement of the veins in the wings of insects.

Neuters Imperfectly developed females of certain social insects (such as ants and bees), which perform all the labours of the community. Hence they are also called 'workers'.

Niche *The ecological niche of an organism includes its habitat, (q.v.), its source of food and shelter, its interaction with other organisms, and all other such factors as define its position in the ecosystem (q.v.).*

Nictitating membrane A semi-transparent membrane, which can be drawn across the eye in birds, reptiles and some mammals, either to moderate the effects of a strong light or to sweep particles of dust from the surface of the eye.

Nucleus *An area within the eukaryotic cell, enclosed by a membrane, and containing the chromosomes and the structures which control cell division.*

Oesophagus The gullet.

Oolitic A great series of Secondary *(Mesozoic)* rocks, so called from the texture of some of its members, which appear to be made up of a mass of small, egg-like, calcareous bodies. *(This type of limestone is present in rocks of all ages but is particularly abundant in the Middle and Upper Jurassic, which were formerly called the Oolitic system.)*

Ova Eggs.

Ovarium or **Ovary (in plants)** The lower part of the pistil or female organs of the flower, containing the ovules or incipient seeds; by growth after the other organs of the flower have fallen, it usually becomes converted into the fruit.

Ovules (of plants) The seeds in the earliest condition.

Pachyderms A group of mammals, so called from their thick skins, and including the elephant, rhinoceros and hippopotamus.

Palaeontology *The study of plants and animals of the past, based primarily on the examination of fossils.*

Parallel evolution *If two forms are derived from a common stock but have independently evolved very similar adaptations to similar ecological niches they are said to show parallel evolution; e.g. Old World monkeys and New World monkeys.*

Parasite An animal or plant living upon or in, and at the expense of, another organism.

Parthenogenesis The production of living organisms from unimpregnated *(unfertilized)* eggs or seeds.

Particulate inheritance *Inheritance by discrete units of hereditary material.*

Pelvis The bony arch to which the hindlimbs of vertebrate animals are articulated.

Phenotype *The observable properties of an organism. The phenotype of an organism includes its external form, its internal structure and its physiological make-up.*

Phenotypic variation *Variation in the phenotype; that is variation which can be observed, either directly, through a microscope, or by biochemical investigations. This term excludes variations in the genes which have no physical expression.*

Photosynthesis *The process whereby light is used as an energy source by living things. It is characteristic of the higher plants, algae, and blue-green algae, (all of which give off oxygen in the process), but is also found in some bacteria.*

Phylum *A major division of the plant or animal kingdom, e.g. phylum Arthropoda.*

Pigment The colouring material produced in the superficial parts of animals. The cells secreting it are called pigment cells.

Pistils The female organs of a flower, which occupy a position in the centre of the other floral organs. The pistil is generally divisible into the ovary, the style and the stigma.

Placentalia or **Placental mammals** See Mammalia.

Pleiotropic *A gene is said to have pleiotropic effects if it controls two or more distinct characters.*

Plutonic rocks Rocks produced by igneous action in the depths of the earth.

Pollen A fine dust produced by the anthers of flowering plants which, by contact with the stigma, effects the fecundation of the seeds. This impregnation is brought about by means of tubes (pollen-tubes) which issue from the pollen-grains. *The pollen grains each contain two male gametes (q.v.).*

Polymorphic Presenting many forms.

Prepotent Having a superiority of power.

Primaries The feathers forming the tip of the wing of a bird.

Processes Projecting portions of bones, usually for the attachment of muscles and ligaments.

Prokaryote *An organism having a simple cell lacking a nucleus; represented by the bacteria and blue-green algae.*

Protein *An organic molecule made up of a chain of amino acids. Proteins are very variable in type, and have a wide range of functions in living organisms. Some are structural, in membranes for example; some have specialized functions, as do the proteins which cause muscle contractions; many act as enzymes, controlling the biochemical processes of the cell.*

Protozoa The lowest great division of the animal kingdom. These animals are composed of a gelatinous material, and show scarcely any trace of distinct organs. The infusoria, foraminifera, and sponges, with some other forms, belong to this division. *(This term is now applied only to single-celled animals, thus excluding the sponges.)*

Pupa (pl. **Pupae**) The second stage in the development of an insect, from which it emerges in the winged reproductive form. In most insects the pupal stage is passed in perfect repose. The chrysalis is the pupal state of butterflies.

Radioactive dating *The use of radioactive substances found naturally in rocks and in organic remains to estimate their age.*

Range The extent of country over which a plant or animal is naturally spread. 'Range in time' expresses the distribution of a species or group through the fossiliferous beds of the earth's crust.

Recent period *The term 'Recent period' was originally used to denote the time since the creation of man and the present order of plants and animals. With the acceptance of evolution this definition became meaningless, but the term Recent is now sometimes used as a synonym for the Holocene or post-glacial epoch, and it is in this sense that Darwin is apparently using it. Note that Darwin considered the Pleistocene epoch to belong to the Tertiary period, so when he speaks of the 'Tertiary and Recent' he is referring to the Cenozoic as a whole.*

Recessive *Of a pair of different genes for the same character the one which will not be expressed if the two different genes are present. See also Dominant.*

Retina The delicate inner coat of the eye, formed by nervous filaments spreading from the optic nerve, and serving for the perception of the impressions produced by light.

Retrogression Backward development. When an animal, as it approaches maturity, becomes less perfectly organised than might be expected from its early stages and known relationships, it is said to undergo a retrograde development.

Reversion *The reappearance in an individual of ancestral characteristics which have not been seen in the intervening generations.*

Rill *A small stream.*

Rodents The gnawing mammals such as the rats, mice and squirrels. They are especially characterised by the possession of a single pair of chisel-like cutting teeth in each jaw, between which and the grinding teeth there is a great gap.

Rubus The bramble genus.

Rudimentary Very imperfectly developed.

Ruminants The group of mammals which ruminate or chew the cud, such as oxen, sheep, and deer. They have divided hoofs, and are destitute of teeth in the upper jaw.

Saltatory *Proceeding by leaps, rather than gradually.*

Sedimentary formations Rocks deposited as sediments from water.

Segments The transverse rings of which the body of an articulate animal or annelid is composed.

Sepals The outermost envelope of an ordinary flower. They are usually green but sometimes brightly coloured.

Sister chromosome *See Homologous chromosomes.*

Somatic *Of the body, as distinct from the reproductive cells.*

Specialisation The setting apart of a particular organ for the performance of a particular function.

Speciation *The process of species formation.*

Species *The modern definition of a species is a group of individuals which can all potentially interbreed, one with another, to produce fertile offspring. The group is genetically isolated from other groups and this genetic isolation allows distinctive features, characteristic of the species, to develop. Note that Darwin avoided defining what was meant by a species, since he wished to argue that there was no sharp dividing line between varieties and species. While modern biologists do give a definition they recognize that species are not fixed but change with time, and that the designation of a group as a species is often somewhat arbitrary.*

Stamens The male organs of flowering plants, standing in a circle within the petals. They usually consist of a filament and an anther, the anther being the essential part in which the pollen is produced.

Sternum The breast-bone.

Stigma The tip of the pistil in flowering plants.

Stipules Small, leafy organs placed at the base of the leaf stalk in many plants.

Style The middle portion of the pistil, which rises like a column from the ovary and supports the stigma at its summit.

Subcutaneous Situated beneath the skin.

Subspecies *A group of individuals within a species having distinctive features, which can interbreed with other members of the species but do not generally do so. See also Variety.*

Suctorial Adapted for sucking.

Sutures (in the skull) The lines of junction of the bones of which the skull is composed.

Tarsus (pl. **Tarsi**) The jointed feet of articulate animals, such as insects.

Taxonomy *The science of classifying living organisms into categories, such as species, genera, families, orders, classes and phyla.*

Teleostean fishes *(Bony fishes)* Fishes of the kind familiar to us in the present day, having the skeleton usually completely ossified and the scales horny.

Trachea The wind-pipe or passage for the admission of air to the lungs.

Trilobites A peculiar extinct group *of marine animals,* somewhat resembling the woodlice in external form, and like some of them capable of rolling themselves up into a ball. Their remains are found only in Palaeozoic rocks, and most abundantly in those of Silurian age.

Typological *The study of characteristic types, which implicitly assumes those types to be fixed.*

Ungulata, Ungulates Hoofed mammals.

Unicellular Consisting of a single cell.

Variety *A group of individuals within a species having one or more distinctive characteristics, but generally able to breed freely with all other members of the species. A variety is less differentiated from other members of the species than is a subspecies.*

Vascular Containing blood-vessels.

Vertebrata or **Vertebrate animals** The highest division of the animal kingdom, so called from the presence in most cases of a backbone *(or vertebral column)* composed of numerous joints or vertebrae, which constitutes the centre of the skeleton and at the same time supports and protects the central parts of the nervous system. *(The fishes, amphibians, reptiles, birds and mammals.)*

Workers See Neuters.

Further Reading

Charles Darwin: A Scientific Biography, Sir Gavin de Beer (Oxford University Press, Oxford, 1958; Doubleday, New York, 1964). The most comprehensive and readable biography available.

Autobiography, Charles Darwin, ed. Francis Darwin (Collier-Macmillan, New York, 1961). A charming autobiography that Darwin wrote for his children.

Autobiography and Selected Letters, Charles Darwin, ed. Francis Darwin (Dover Publications, New York).

Charles Darwin and Thomas Henry Huxley: Autobiographies, ed. Sir Gavin de Beer (Oxford English Memoirs and Travel Series, Oxford University Press, Oxford, 1974).

Apes, Angels and Victorians: a joint biography of Darwin and Huxley, William Irvine (Meridian, New York, 1965).

Darwin and his Flowers: The Key to Natural Selection, Mea Allen (Faber and Faber, London, 1977). A delightful, illustrated biography centred around Darwin's botanical studies.

Darwin and the Beagle, Alan Moorehead (Hamish Hamilton, London, 1969; Penguin, London, 1971; Harper & Row, New York, 1969). A highly illustrated biography of Darwin with the emphasis on his time aboard the *Beagle*.

The Voyage of the Beagle, Charles Darwin (Everyman Library, Dent, London). Darwin's very readable, personal account of the five-year voyage.

Darwin's Century: Evolution and the men who discovered it, Loren Eiseley (Gollancz, London, 1959; Doubleday, New York, 1958). A good account of nineteenth-century evolutionary thought.

Ever Since Darwin: Reflections in Natural History, Stephen Jay Gould (Burnett Books and Andre Deutsch, London, 1978; Norton, New York, 1977). An entertaining and stimulating collection of essays on a wide range of topics, both historical and contemporary, relating to Darwin and his theory.

Evolution, Colin Patterson (British Museum, Natural History, 1978). A clear, simple, illustrated account of modern knowledge about evolution.

The Meaning of Evolution, George Gaylord Simpson (Yale University Press, New Haven, Connecticut, 1967). A fuller survey, with the emphasis on palaeontology.

The Theory of Evolution, John Maynard Smith (Penguin, London, 1976). A detailed, comprehensive, and yet highly readable account of all aspects of evolutionary theory.

Natural Selection and Heredity, Philip Sheppard (University Library, Hutchinson Educational, London, 1975). A study of natural selection in the light of modern genetics. Many of the examples mentioned in the introduction (*eg* industrial melanism in the peppered moth) are described in full detail here.

Origins: What New Discoveries Reveal About the Emergence of Our Species and its Possible Future, Richard Leakey and Roger Lewin (Macdonald and Janes, London, 1977; Dutton, New York, 1977). A beautifully illustrated account of current research into the origin of man.

Sociobiology: The New Synthesis, Edward O. Wilson (Harvard University Press, Cambridge, Massachusetts, 1975). A comprehensive survey of what is known about behaviour and instincts; the last chapter introduces the ideas of sociobiology.

The Selfish Gene, Richard Dawkins (Oxford University Press, Oxford, 1976; Paladin, St. Albans, Hertfordshire, 1978). A popular exposition of sociobiological thought.

Illustration Acknowledgments

97 British Museum (Natural History)
98 Popperfoto
101 Daphne Machin-Goodall
105 British Museum (Natural History)
106 Peter Ward/Bruce Coleman
107 *top*: Ardea, London
   bottom: Charlie Ott/Bruce Coleman
108 Gunter Ziesler/Bruce Coleman
109 photo: Derek Witty. Courtesy of The London
   Library
110 diagrams: Anthony Maynard
   photos, *top*: Laurie Collard
   middle: N. A. Callow/NHPA
   bottom: Jane Burton/Bruce Coleman
113 Anthony Maynard
114 Jane Burton/Bruce Coleman
117 *top*: J. M. Start/Robert Harding Associates
   bottom: O. Luz/ZEFA
118 Syndics of Cambridge University Library
119 Derek Witty
120 Pat Morris/Ardea, London
124 Masood Qureshi/Bruce Coleman
125 *top*: A–Z collection
   bottom: Dr David Corke/ZEFA
127 D. Baglin/NHPA
128 Stephen Dalton/Bruce Coleman
   S. C. Bisserôt/Bruce Coleman
132 *top*: John Mason/Ardea, London
   bottom: Heather Angel
133 Syndics of Cambridge University Library
134 J. P. Ferrero/Ardea, London
136 Aquila Photographics
137 photo: Wayne Lankinen/Bruce Coleman
   artwork: Hilary Burn. From photograph,
   copyright: John Topham Picture Library
139 *top*: Gunter Ziesler/Bruce Coleman
   bottom: A. J. Deane/Bruce Coleman
141 Derek Witty
143 S. C. Bisserôt/John Topham Picture Library
145 Eric Crichton/Bruce Coleman
147 Jen and Des Bartlett/Bruce Coleman
149 Derek Witty
152-3 Aerofilms Ltd
154 Walter Rawlings/Robert Harding Associates
155 Popperfoto
157 Robert Harding Associates

159 Anthony Maynard
161 Popperfoto
162 Laurie Collard
163 Heather Angel
164 Derek Witty
168-9 Anthony Maynard
   maps: Eugene Fleury
170 The Mansell Collection
172 map: Eugene Fleury
   animals: Anthony Maynard
175 Syndics of Cambridge University Library
176 *top*: John Markham/Bruce Coleman
   bottom left: Michael Freeman/Bruce Coleman
   bottom left: Francisco Erize/Bruce Coleman
181 photo: Radio Times Hulton Picture Library
   map: Eugene Fleury
184 *top*: Heather Angel
   bottom: Ake Lindau/Ardea, London
186-7 Aerofilms Ltd
188-9 Nicholas Hall
190 Derek Witty
192 Ardea, London
193 *top*: Heather Angel
   bottom: Densey Clyne
196-7 Eugene Fleury
199 Anthony Maynard
200 Heather Angel
201 Heather Angel
202 Derek Witty
203 Derek Witty
207 Syndics of Cambridge University Library
208 Anthony Maynard
209 British Museum (Natural History)
211 Anthony Maynard
212 British Museum (Natural History)
213 British Library
214 Anthony Maynard
217 Mary Evans Picture Library
220 Dr Walker/NHPA
221 Syndics of Cambridge University Library
223 British Library

The publishers would particularly like to thank the
following people for their help: Heather Angel, Jen
and Des Bartlett, Daphne Machin-Goodall, and
Richard Moody of the RIDA Photo Library.

Index

Formation of the earth's crust

Origin of life : bacteria first appear

| 4500 million years ago | 4000 million years ago | 3500 million years ago | 3000 million years ago | 2500 million years ago |

Most invertebrate groups appear in fossil record for first time

Trilobites numerous

Fish first appear

Insects first appear

Amphibians first appear

Reptiles first appear

PALAEOZOIC

| PRECAMBRIAN | CAMBRIAN | ORDOVICIAN | SILURIAN | DEVONIAN | CARBONIFEROUS |

| 600 million years ago | 500 million years ago | 400 million years ago | 300 million years ago |